D0968930

SCIENTIFIC PROGRESS

SYNTHESE LIBRARY

STUDIES IN EPISTEMOLOGY,

LOGIC, METHODOLOGY, AND PHILOSOPHY OF SCIENCE

VOLUME 153

CRAIG DILWORTH

Department of Philosophy, Uppsala University, Uppsala, Sweden

SCIENTIFIC PROGRESS

A Study Concerning the Nature of the Relation Between Successive Scientific Theories

D. REIDEL PUBLISHING COMPANY

DORDRECHT : HOLLAND / BOSTON : U.S.A.

LONDON : ENGLAND

Library of Congress Cataloging in Publication Data

CIP

Main entry under title:

Dilworth, Craig.
Scientific Progress.

(Synthese Library; v. 153)
Bibliography: p.
Includes index.
1. Science–Philosophy. 2. Science–Methodology. I. Title.
Q175.D6342 501 81-13814
ISBN 90-277-1311-1 AACR 2

Published by D. Reidel Publishing Company
P.O. Box 17, 3300 AA Dordrecht, Holland

Sold and distributed in the U.S.A. and Canada
by Kluwer Boston Inc.,
190 Old Derby Street, Hingham, MA 02043, U.S.A.

In all other countries, sold and distributed
by Kluwer Academic Publishers Group
P.O. Box 322, 3300 AH Dordrecht, Holland

D. Reidel Publishing Company is a member of the Kluwer Group.

Printed in Sweden by Almqvist & Wiksell, Uppsala, 1981

ACKNOWLEDGEMENTS

This study has been presented in a number of versions since its central ideas first appeared in a short essay "Incommensurability and Scientific Progress" in 1975. Since that time it has received the helpful criticism of very many people, and to all of them I here express my thanks. Lennart Nordenfelt has discussed with me in detail versions appearing in 1976 and 1977. Those whose contributions are more recent include Rainer Carls, Paul Feyerabend, Mats Furberg, Lars Hertzberg, and Dag Prawitz, each of whom has made valuable suggestions concerning my 1978 paper "On the Nature of the Relation Between Successive Scientific Theories", from which the last six chapters of the present work have been developed. More specialized comments have been offered by Staffan Nilsson (Chapter 10), and by Peter Gärdenfors and Włodzimierz Rabinowicz (Chapter 11). And for reading and commenting on the whole of this work just before it went to press, I express my gratitude to Ingvar Johansson and Giovanni Sommaruga.

Very special thanks are due to Prof. Stig Kanger, for his support and guidance during my years as a graduate student at Uppsala, and to Prof. Evandro Agazzi, who has done everything possible to help see this work through to completion.

C. D.

TABLE OF CONTENTS

INTRODUCTION

For the philosopher interested in the idea of objective knowledge of the real world, the nature of science is of special importance, for science, and more particularly physics, is today considered to be paradigmatic in its affording of such knowledge. And no understanding of science is complete until it includes an appreciation of the nature of the relation between successive scientific theories—that is, until it includes a conception of scientific progress.

Now it might be suggested by some that there are a variety of ways in which science progresses, or that there are a number of different notions of scientific progress, not all of which concern the relation between successive scientific theories. For example, it may be thought that science progresses through the application of scientific method to areas where it has not previously been applied, or, through the development of individual theories. However, it is here suggested that the application of the methods of science to new areas does not concern forward progress so much as lateral expansion, and that the provision of a conception of how individual theories develop would lack the generality expected of an account concerning the progress of science itself.

In considering the nature of scientific progress through theory change, a particular feature of the relation between theories presents itself as requiring explanation. This feature is the competition or rivalry that exists between successive theories in their attempts to explain certain aspects of reality. We note then that an adequate account of scientific progress should include a conception of the conflict that arises in the case of successive scientific theories.

In its treatment of the notion of scientific progress, this study begins with a critical analysis of the logical empiricist and Popperian conceptions of the nature of the relation between successive theories, and the basis from which these conceptions are derived. The analysis is structured via the reconstruction of the Empiricist and Popperian conceptions in terms of the covering-law model of explanation. (As employed in the reconstruction, the model is more simply termed: the Deductive Model.) This reconstruction is intended to give a precise clarification of the Empiricist and Popperian views, revealing both their capabilities and their limitations.

The major criticisms based on the reconstruction are of three kinds. The first of these concerns the inability of the Deductive Model, as employed by the Empiricists and Popper respectively, to *formulate* conceptions of important aspects of science. The most important of these criticisms are that the Empiricist view affords no notion of theory conflict, and that the Popperian view fails to provide a notion of scientific progress. The second type of criticism concerns the problem of *applying* the Deductive Model to actual science. Thus, for example, it will be shown that while the Empiricists have given us a notion of scientific progress (as involving deductive subsumption), actual scientific advance does not take this form. The third kind of criticism, directed mainly at Popper, suggests that a number of claims considered to be integral to his view are actually quite *ad hoc*, in that they are not at all suggested by the Deductive Model, from which his conception of science has earlier been derived.

The study then moves on to consider the important claims of Thomas Kuhn and Paul Feyerabend, that in certain cases succeeding theories might well be 'incommensurable' with their predecessors. These claims, in their negative aspect, are viewed as essentially being criticisms of the second sort mentioned above. That is, they are seen to concern the applicability of the Empiricist and Popperian conceptions of the relation between theories, and to suggest the relinquishment of the model underlying these views. And, in their positive aspect, they are taken to suggest that in actual science theories are often related in the same sort of way as are the different aspects of a gestalt switch diagram.

Following this lead, a model which is fundamentally different from that of the Empiricists and Popperians is introduced. This model—the Gestalt Model—is intended to provide a positive understanding of incommensurability, and to afford notions both of conflict and of progress.

On the basis of this model an alternative conception of science and scientific progress is then presented, in which scientific theories are seen not to be entities of the sort which are either true or false, but to be structures which are more or less applicable depending on the results of certain measurements.

Following this, the conception of the present study is developed further in the context of its application to the kinetic theory of gases. In this development the role of *models* in theoretical science, while not treated by the Empiricist and Popperian views, becomes of central importance.

A critique is then made of the set-theoretic or structuralist conception of science, in which a notion of model also plays a role. An examination of the reconstruction of Newtonian particle mechanics in terms of intuitive set theory, and the attempted extension of these methods to the case of theory change, finds them not to have provided adequate conceptions either of theory conflict or of scientific progress.

Finally, the conception of the present study is applied to the views of Newton, Kepler, and Galileo concerning the motions of material bodies. Here the opportunity is also taken to compare the major conceptions of science and scientific progress in this context; and it is hoped that this comparison helps further to establish the relative superiority of the view offered in the present study.

CHAPTER 1

THE DEDUCTIVE MODEL

As will be shown in this study, the Deductive Model constitutes the formal basis upon which both the logical empiricist and Popperian conceptions of science and scientific progress are built. It is here introduced in its most familiar form: as a model of explanation and prediction.

1. THE DEDUCTIVE MODEL AS A MODEL OF EXPLANATION

Popper formulates the model in his *Logic of Scientific Discovery*[1] as a model of causal explanation consisting in universal statements and singular statements, the conjunction of which entails some particular prediction. The model takes its more classic form as the covering-law model of deductive-nomological explanation in an article by Carl Hempel and Paul Oppenheim entitled "Studies in the Logic of Explanation" (1948). It is there summarized in the following sort of schema:

(1)

Logical deduction	$L_1, L_2, \ldots, L_r$	General Laws	Explanans
	$C_1, C_2, \ldots, C_k$	Statements of antecedent conditions	
	$\rightarrow E$	Description of the empirical phenomenon to be explained	Explanandum

Taken to its simplest extreme, the model may be schematized as follows:[2]

(2) $A \vdash B$,

where "A" is to denote the conjunction of general laws and statements of antecedent or initial conditions, and "B" is to denote the

[1] Popper (1934), pp. 59f.

[2] Cf. Popper (1934), p. 76.

explanandum. The deductive relation is to go from *A* to *B* so that the truth of *A* is sufficient for the truth of *B*, and the truth of *B* is necessary for that of *A*.

The model may also be presented in a slightly more complicated form thus:

(3) $L \,\&\, C \vdash E.$

Here "*L*" denotes the conjunction of general laws, "*C*" the conjunction of statements of initial conditions, and "*E*" the explanandum.

In the model general laws are taken to be unrestricted universal statements; and statements of initial conditions and the explanandum may be conceived as restricted or specific statements. (Unrestricted statements are to be applicable at any place at any time, while restricted statements are to relate only to specific times and places. For example, on this line of thinking 'All swans are white' and 'There exists a black swan' are unrestricted; and 'This swan is black' is restricted.)

The formally deductive nature of the model can be captured by its being formulated directly in terms of the first-order predicate calculus. Here only one law and one statement of initial conditions will be taken to be present:

(4)	L	$(x)(Fx \rightarrow Gx)$	(unrestricted)
	C	Fa	(restricted)
thus,	E	Ga	(restricted).

And an example of the application of the above might go as follows:

(5)	$(x)(Fx \rightarrow Gx)$	All copper expands when heated
	Fa	This is copper being heated
thus,	Ga	This (copper) expands.

Here "*F*" stands for: 'is copper being heated', and "*G*" stands for: 'expands'.

It is of some interest to note that the model, as given in (4) and (5) above, not only bears a close affinity to the Aristotelian syllogism,[3]

[3] Cf. *An. Pr.* 26ᵃ 24–27: "Let all *B* be *A* and some *C* be *B*. Then if 'predicated of all' means [that no instance of the subject can be found of which *B* cannot be asserted], it is necessary that some *C* is *A*. ... So there will be a perfect syllogism."

but is in fact a Stoic syllogism of the same form as: All men are mortal; Socrates is a man; thus, Socrates is mortal.[4]

As presented by Popper and by Hempel and Oppenheim the model can function in two different ways: it can serve both as a model of explanation and as a model of prediction. In applying it as a model of explanation it is supposed that those receiving the explanation are aware of the truth of *E*, and are being informed that *L* and *C* are the case. In its application to prediction, *L* is assumed true, and, following the establishment of *C*, the truth value of *E* is to be empirically determined.[5]

2. A CRITICISM OF THE MODEL AS A MODEL OF EXPLANATION

As presented above, the Deductive Model is to afford the *linguistic form* that a line of reasoning ought ideally to have in order to count as an explanation or prediction. The model has been the recipient of a number of criticisms, most of which concern either the existence of seemingly adequate explanations (or predictions) not having the form suggested by the model (e.g. teleological explanations), or the fact that certain lines of reasoning have the deductive form, and yet are not explanations (e.g. conventional generalizations).

All the same, it is believed by many that the rigorous explanations made in deterministic branches of science do in fact have the deductive form, and that in this way it constitutes a sort of ideal, deserving of emulation. But it is suggested here that explanations in science do not have this form, and that those cases to which the model actually has been applied do not constitute instances of explanation.

If we consider the (paradigmatic) example given above, and suppose someone to witness the expansion of some particular material under certain conditions and to seek a scientific explanation of this phenomenon, it is not at all clear that his being told that the material is copper being heated, and that all copper expands when heated, would provide him with what he is seeking. In other words, his knowing that

[4] Concerning the Stoic origin of syllogisms having this form, see Bocheński (1961), p. 232 &n.

[5] Cf. e.g. Hempel (1962), pp. 118–119.

all copper behaves in this way under these conditions does not tell him *why* this particular piece does so; it only tells him that, in being constituted of copper, if this piece were replaced with another, also constituted of copper, the substituted piece would behave in the same way. And this would still leave him without an explanation as to why this material, which he now knows to be copper being heated, should expand under these circumstances.[6]

This problem will be seen later in this study to stem from the fact that explanations are here viewed as being based on scientific laws, rather than on theories. But of greater interest at this point is the fact that, as will now be shown, not only can both the Empiricist and Popperian conceptions of science and scientific progress be derived from the Deductive Model, but both their capabilities and limitations are bound to it.[7]

---

[6] For a similar criticism, see Scriven (1962), p. 203; concerning the applicability of the Deductive Model to the case of laws being explained by higher-level theories, see Chapter 4 below.

[7] The present reconstruction of the Empiricist and Popperian conceptions in terms of the Deductive Model may be seen as a presentation of what has recently come to be called 'the statement view'. Cf. e.g. Stegmüller (1973), p. 2, and Feyerabend (1977), p. 351.

CHAPTER 2

THE BASIS OF THE LOGICAL EMPIRICIST CONCEPTION OF SCIENCE

1. VERIFIABILITY

Logical empiricism is an outgrowth of logical positivism, in which the verifiability principle was put forward as a criterion for distinguishing meaningful statements from meaningless pseudo-statements. For logical positivism, if any proposition or statement were not in principle conclusively verifiable by experience, it was to be considered meaningless, or, at best, tautological. Along this line then it was intended that meaningful statements include the pronouncements of science, while excluding those of metaphysics, ethics, and theology.

With the realization that on this criterion scientific laws would themselves be meaningless, a first step toward the logical empiricist position was taken by extending the status of meaningfulness to any proposition from which an empirically verifiable proposition could be logically derived. In such derivations, the meaningfulness of the consequent was to imply that of the antecedent.

But this view too suffered problems, a major one of which centres on the fact that no universal statement or law by itself entails an observation statement. What more is required is a statement of the conditions under which the observation is being made.

In this way then alterations in the Positivist criterion of meaningfulness give rise to the basis of the Empiricist conception of science, in which scientific laws, in conjunction with statements of initial conditions, are to entail particular observation statements. Here we see that lying behind these developments is the conception of laws, statements of conditions, and observation statements as having the form suggested by the Deductive Model: $(L \& C) \vdash E$.

Unfortunately, for logical reasons the above attempt to include scientific laws among meaningful assertions while excluding metaphysical sorts of statements has proved unsuccessful.[1] But more important here is the introduction of the Deductive Model at the basis of the Empiricist conception of science.

2. INDUCTION AND CONFIRMATION

As well as affording a structure for explanation and prediction, and for the above criterion of meaningfulness, the Deductive Model can be seen to form the basis of the Empiricist conception of induction. Where in applying the model to explanation the starting point taken is the truth of the explanandum, and in the case of prediction it is the truth of the laws and statements of conditions, in its application to induction one takes the statements of conditions and explanandum, i.e. (C & E), or (Fa & Ga), to be true. Thus while we should not say that scientific laws, as conceived on the model, are logically derivable from statements of conditions and explananda, we may say that they can be related to the latter by means of induction.

A point not always recognized in discussions concerning (empirical) induction is that the notion itself has two distinct applications. In one sense, induction may be thought of as a possible means by which we come to realize that there exist certain regularities in nature on the basis of an acquaintance with their instances. In its other sense, induction may be considered the method employed to afford rational support for the claim that some particular regularity does in fact exist. As conceived on the Deductive Model, both of these senses are fundamental to logical empiricism, the latter (called 'confirmation') defining its position in the context of justification, and the former (here termed simply 'induction') in the context of discovery.

The main problem with the Empiricist conception of *induction* as being the means by which new laws are discovered, as has been noted by others, is that it provides no hint as to why attention is focussed on certain particular phenomena as providing the basis from which the inductive step is taken. The scientist seldom simply amasses quantities of data, sifting through them hoping to find a regularity. Rather,

---

[1] For a discussion of this problem see Ch. 4 of Hempel (1965). For a presentation of the view being reconstructed in the present chapter, see e.g. Ayer (1936).

he usually works in the context of some theory which, as will be discussed later in this study, is not itself a regularity of the same sort as that being sought.

In various forms the problem of *confirmation* has received a great deal of attention in Empiricist writings. The heart of this problem lies in the fact that the truth of the conclusion of a logical deduction does not imply the truth of the premisses. In terms of the Deductive Model the problem is that the truth of the explanandum and statements of conditions, i.e. the truth of statements of the form (Fa & Ga), does not establish the truth of the law. And not only this, but since the law is conceived as an unrestricted universal statement, no one finite number of true statements of the above form provides any more logical support for it, or makes its truth any more probable, than does any other. But it may be pointed out that, if we do grant scientific laws as conforming to the model, then this problem of induction is not a problem for the Empiricist, but for the scientist, for all that is demanded of the Empiricist is that he provide a conception of science as it is actually practiced.

But then it may be asked whether scientific laws do in fact have the form suggested by the Deductive Model. An examination of the nature of scientific laws, at least in the exact sciences, reveals that, rather than being expressed by universal statements having a truth value, they are most often expressed as *equations* suggesting a numerical relationship among the values of certain parameters. And, where on the Deductive Model it is difficult to see how a universal statement discovered to be false might nevertheless continue to function as the expression of a law of nature, in science we find that laws expressed by equations are often retained even when it is realized that they have only a limited range of application.[2]

While a positive account of the nature of scientific laws which is in keeping with the above observations will be given later, for present

[2] It is perhaps worth mentioning that it is not the intention of the present study to suggest that Empiricists are unaware of such facts as these. But the development of the Empiricist philosophy of science in the earlier works of such authors as Carnap, Hempel, and Ayer does not give such facts a place. And to the extent that the Empiricists have provided an account or explanation of science, it is here intended that the essence of that account, i.e. its formal basis, be captured in the present reconstruction. These same remarks apply, *mutatis mutandis*, to Popper and Lakatos.

purposes it suffices to point out that the Empiricist conception of science may be seen as being formally based on the Deductive Model, and that in this way it thus begins with a conception of scientific laws, rather than theories. In the next chapter the basis of the Popperian conception of science will be treated, and it too will be found to rest on the Deductive Model.

CHAPTER 3

THE BASIS OF THE POPPERIAN CONCEPTION OF SCIENCE

1. FALSIFIABILITY

The considerations of the previous chapter indicate that the Positivist and Empiricist views can be seen as attempting to demarcate (meaningful) science from (meaningless) non-science on the basis of verifiability and confirmability respectively. Popper's demarcation between science and non-science, on the other hand, is on the basis of falsifiability. For Popper, if there is no conceivable way that a statement can be shown to be false, while it might still be considered meaningful, it is not scientific but 'metaphysical'.

Seen most simply, on the Empiricist conception the confirmation of scientific laws consists in the verification of observation statements entailed by them. On Popper's view, in its simplest form, laws or theories may be falsified via the determination of the truth of observation statements that contradict them. Thus where we can represent the Empiricist conception by $A \vdash B$, where A is to include a universal law, and B observational evidence, Popper's conception can here be represented by the formally equivalent:

(6) $\neg B \vdash \neg A$.

Here we see that the Popperian view lays stress on the idea that, while no amount of true observation statements of the sort B could verify the universal statement in A, the truth of one observation statement $\neg B$ should suffice to falsify A.[1]

The fact that at this primitive stage the schematization of Popper's view is formally equivalent to that of the Empiricist conception is worthy of note, for it suggests that the difference between the bases of the two views is more one of emphasis than of substance.[2] Where

[1] On this point, see Feyerabend (1974), p. 499.

[2] Cf. a similar remark by Carnap cited in Popper (1962), p. 254 n.

the Empiricist directs himself to the problem of what justifies our believing certain general claims of science to be valid, Popper points to a criterion—capable of being formulated within the Empiricist conception—which should suffice to show them to be invalid. Of course the refutability of general claims in science was generally recognized before Popper made falsifiability his criterion of demarcation,[3] and, as will be seen below, Popper's main contribution beyond his demarcation criterion is his attempt to develop this idea in terms representable by the Deductive Model.

The basis of the Popperian view as outlined to this point, in emphasizing the falsifiability of general or universal claims, can be seen to have two serious shortcomings in comparison with a similarly simple presentation of the Empiricist view. It affords a conception neither of the discovery of new laws (context of discovery), nor of the support of claims that certain laws exist (context of justification).[4] If it is viewed as providing a conception of discovery, such discovery is the discovery of mistakes; and granting that it affords a conception of justification (in a broad sense), such justification is the justification one might have in saying that something is wrong.

2. BASIC STATEMENTS AND BACKGROUND KNOWLEDGE

A first step in rendering Popper's conception more sophisticated parallels a move made by the Empiricists in their development of the confirmability criterion of meaningfulness. It is the recognition that —in keeping with the conception of laws suggested by the Deductive Model—a universal statement entails an observation statement only when the former is conjoined with certain statements of initial conditions. In the case of the Empiricist view this may be schematized by: $(L \,\&\, C) \vdash E$. In Popper's case the formulation is again equivalent,

---

[3] Cf. e.g. Poincaré (1903), pp. 150ff., Duhem (1906), pp. 180ff., and Campbell (1920), pp. 109 & 131.

[4] Lakatos would almost certainly have disagreed with this. On p. 375 of his (1968) he claims Popper to have focussed attention on the problem of the discovery of hypotheses, when he has in fact focussed attention on their refutation. But Lakatos' conception of the 'logic of discovery' is rather unusual—he sees it as the discipline of the rational *appraisal* of theories: in this regard see his (1970), p. 115.

but in keeping with the $\neg B \vdash \neg A$ schematization in (6), it might first be presented as follows:

(7) $$\neg E \vdash \neg (L \,\&\, C).$$

This formulation of the basis of Popper's philosophy of science in terms of the Deductive Model makes it clear that his notion of falsification is not so straightforward as one might have hoped. Here, where we begin with the 'basic statement' $\neg E$,[5] we find that it does not entail the negation of the law or theory L, but rather entails the negation of the conjunction of L and the statement(s) of conditions C. Thus the determination of the truth of $\neg E$ would not suffice to falsify L.[6]

In order to obtain a situation in which L *is* falsified, Popper employs a line of thought that can best be represented by:

(8) $$(C \,\&\, \neg E) \vdash \neg L.$$

In this formulation, which is still formally equivalent to the basis of the Empiricist conception—i.e. to the Deductive Model—$(C \,\&\, \neg E)$, or, in the predicate calculus $(Fa \,\&\, \neg Ga)$, is the 'falsifying basic statement' deductively subsuming the negation of the law.[7] But it is obvious that this does not avoid the problem, for it is still the case that, just as the Empiricist conception requires the truth (or confirmation) of C in order to confirm L, Popper requires the truth of C in order to falsify L.

Now, where the Empiricists might want to say that statements of the sort C, e.g. 'This is copper being heated', are capable of being observationally verified (or at least confirmed), Popper, in a move away from Empiricism (Positivism), argues that since such statements contain universal notions like 'copper', which themselves are based on certain theoretical presuppositions, they *cannot* be verified.[8] He suggests, in fact, that like general laws themselves, such statements can be falsified, but can be neither confirmed nor in any way estab-

---

[5] Cf. Popper (1959), p. 85 n.

[6] For a similar point, see Duhem (1906), p. 185.

[7] Cf. e.g. Popper (1934), pp. 102 & 127, and (1959), p. 85 n.

[8] Popper (1934), pp. 94–95, (1959), pp. 423–424 &n. It may be noted that Carnap also adopts this stance in his (1936), pp. 425 ff. In this regard cf. also Campbell (1920), p. 43.

lished as true. Thus for Popper, in a situation such as that depicted by (8), statements of the sort C are to be 'background knowledge', which for the purpose of falsifying L are to be *tentatively* accepted as unproblematic.[9] But if we ask why they (or the basic statements containing them) are to be so accepted, we are told that in certain cases it is because they have survived similar attempts at falsification.[10] On Popper's reasoning, however, this leads to an infinite regress which is ultimately ended only by the consensus of scientists. But the basing of falsifications on consensus, while perhaps a fair procedure, is hardly an objective one, and does not suffice to distinguish science from other pursuits.

In any case, the fact that Popper suggests that statements C cannot be established as true, for whatever reason, means that general statements cannot be established as false.[11] And this in turn means that Popper cannot claim to have provided a criterion for distinguishing science from non-science on the basis of the falsifiability of the claims made by the former.[12]

3. CORROBORATION, SEVERITY OF TESTS, AND THE FALSIFIABILITY OF THE EMPIRICAL BASIS

One of the advantages Popper claims for his view is that it solves the problem of induction. As was mentioned in the previous chapter, the notion of induction may be seen to have two distinct applications: one to the discovery of new laws, and one to the justification of claims that a particular law has in fact been found. Popper focusses on the latter application, and believes himself to have solved the problem of induction through suggesting that our reason for accepting certain theories in science is not because they have been established as true, but because they have survived (severe) attempts at falsification.[13]

---

[9] Cf. e.g. Popper (1962), pp. 390f.

[10] Popper (1934), p. 104.

[11] For a similar criticism see Deutscher (1968), p. 280; see also Duhem (1906), p. 185.

[12] Note that the present criticism does not preclude some sort of testability criterion of demarcation, but relates directly to Popper's *falsifiability* criterion as it is to be understood in the context of his conception of scientific laws.

[13] Cf. e.g. Popper (1973), p. 8.

And theories that have survived such tests are, as a result, to be considered 'corroborated'.

Of course few people, if any, would claim that the determination of true consequences of an empirical law would *establish* the law as true; and, as it has been presented in the previous chapter, the Empiricist problem of confirmation concerns rather the formulation of a conception in which, roughly speaking, the determination of the truth of consequences of the law would add *support* to the claim that the law is true.[14] Nevertheless, Popper for the most part directs his critical arguments in this context against whomever it might be that believes that laws can be completely verified.[15]

However, at one point Popper does treat of the case where the truth of the consequences of a law may be thought to add confirmatory support to the law. His argument here can be clearly presented in terms of the Deductive Model. If we assume a law to have the form $(x)(Fx \rightarrow Gx)$, then for any object a, the law implies that $(Fa \rightarrow Ga)$. Statements having the latter form Popper calls 'instantial statements', and suggests that since they, or their logical equivalents $(\neg Fa \vee Ga)$ and $\neg(Fa \;\&\; \neg Ga)$, are verified wherever $\neg Fa$ is the case—i.e. almost everywhere—"these instantial statements *cannot play the role of test statements* (or of potential falsifiers) which is precisely the role which basic statements are supposed to play."[16]

Popper is perhaps here overlooking the fact that these 'instantial statements' are precisely the statements which must be determined true in order to falsify basic statements; and his argument that such statements cannot function as test statements in the context of universal laws applies just as well in this other context, and results in Popper's implicitly denying scientific status to his own basic statements. But, in any case, if we are to interpret Popper's argument as a criticism of the Empiricist notion of confirmation we see that it is wide of the mark, for, as has been pointed out in the previous chapter, on the Empiricist view it is not statements of the form $(Fa \rightarrow Ga)$ that should be thought to confirm a law, but statements of the form $(Fa \;\&$

[14] Cf. Hempel (1966), pp. 116f.

[15] Popper's only reference in this regard is to an article written by Gilbert Ryle in 1937: cf. Popper (1973), p. 9.

[16] Popper (1959), p. 101 n.

Ga).[17] And the verification of such statements is no more problematic than is the verification of Popper's basic statements.

But the problem of confirmation, as described in the previous chapter, still remains, and so we might investigate whether Popper has succeeded in solving or avoiding it via the employment of his own notion of corroboration. What is being required of him here then is the presentation of a conception not involving the notion of induction, in terms of which we can understand how the passing of (severe) tests should make more reasonable the acceptance of a particular scientific law.

In keeping with Popper,[18] his definition of corroboration[19] may also be presented in terms of the Deductive Model:

$$(9) \qquad C(L, E, C) = \frac{P(E, LC) - P(E, C)}{P(E, LC) - P(LE, C) + P(E, C)}.$$

Here "$C(L, E, C)$" stands for: "the degree of corroboration of law L by explananda E, in the presence of conditions (background knowledge) C"; and, for example, "$P(E, LC)$" means: 'the relative probability of the explananda E, given the truth of the conjunction of L and C'.

Disallowing induction, the above suggests that the degree of corroboration of all laws would be the same: $1 - 0/1 - 0 + 0 = 1$. (This is not surprising considering that Popper sees no possibility of attaching numerical values other than 0 and 1 to his measures of probability.)[20] But what this means is that Popper has not succeeded in providing a metric for degree of corroboration.[21]

However, all that is being asked of Popper here is that he provide a non-inductive conception of how the passing of tests should corroborate a theory, and so his failure to provide a metric may be overlooked, and we might investigate instead the ideas lying behind his definition, to see if such a conception may be found there. Thus we

[17] We note that it is normally statements of this latter form which are called instantial, and that what Popper calls instantial statements are actually hypothetical statements.
[18] Cf. Popper (1962), pp. 390 f.
[19] Popper (1959), p. 400 n.
[20] Cf. Popper (1962), p. 397.
[21] In this regard cf. Lakatos (1968), p. 396.

note that if it were not for the middle term in the divisor (which Popper especially inserts in order to satisfy certain intuitive desiderata), the degree of corroboration of L would depend solely on the value of $P(E, C)$;[22] i.e., as the probability of E, given background knowledge C, rises, so should the degree of corroboration of L.

Underlying Popper's thinking here is that as each test of L is passed, the results of the test are to be added to the background knowledge; thus, the more tests L passes, the greater is the probability that it will pass the next one. In this way too, on Popper's view, the *severity* of each successive test is to decline. In Popper's own words:

> [I]f a theory stands up to many such tests, then, owing to the incorporation of the results of our tests into our background knowledge, there may be, after a time, no places left where (in the light of our new background knowledge) counter examples can with a high probability be expected to occur. But this means that the degree of severity of our test declines.[23]

And, as the severity of tests declines and the probability of E given C rises, so too rises the degree of corroboration of L.

However, while this line of thinking seems intuitively reasonable, it can be of no use to Popper, since it is inductive, and, consequently, so are Popper's concepts of corroboration and severity of tests. Not only this, but if we stop to consider the way in which Popper conceives of a theory being tested we see that his notion of corroboration, or the passing of tests, is essentially identical to the Empiricist notion of confirmation. For the Empiricist the determination of the truth of a statement of the form $(Fa \& Ga)$ should provide confirmative support for the truth of a law of the form $(x)(Fx \rightarrow Gx)$. For Popper, such laws are tested by attempting to discover acceptable falsifying basic statements of the form $(Fa \& \neg Ga)$. Both notions require the determination of the truth of Fa, and if it is also found that $\neg Ga$ is the case, then the law is falsified. But if its negation, Ga, is the case then, on the Empiricist view, the law has been confirmed,

[22] On this point cf. Lakatos (1968), p. 410 n.

[23] Popper (1962), p. 240. Popper's notion of severity of tests, given here, ought not be confused with his notion of testability, which is essentially the same as his concept of content (though Popper himself sometimes conflates these two notions: cf. e.g. Popper (1959), p. 374).

and, on Popper's view, it has passed it test and has consequently been corroborated.

As a result of the above considerations then we see that Popper is not at all warranted in claiming to have solved the problem of induction, even when it is limited to the context of justification. His notions of corroboration and severity of tests are both inductive, and do not differ in any significant way from the Empiricist conception of (degree of) confirmation; and his conception of the testing of a theory is also essentially the same as that of the Empiricists.

But if we question whether scientific laws actually do have the form suggested by the Deductive Model, as has been done at the end of the previous chapter, then the sorts of questions treated in this chapter, framed in terms of the model, become mainly of academic interest. That this is so is even more evident when it is realized that Popper makes no distinction between scientific laws and theories, taking both to have the form suggested by the model.[24]

It is hoped that the efforts of the last two chapters have succeeded in showing both Popper's conception and that of the Empiricists to be formally based on the Deductive Model. Both views take scientific laws to be universal statements having a truth value, and the extent to which we know this truth value is to be based on our knowledge of the truth values of certain explananda and statements of conditions. Given true statements of conditions (assuming their truth determinable) and a false explanandum, the law is considered false, whereas true statements of conditions and true explananda are to confirm or corroborate the law. And the essential difference between the Empiricist and Popperian views as developed thus far may be seen to lie in the Empiricists' normally applying the model to cases where the explanandum may be considered true, where Popper usually applies it to cases in which the explanandum should be false. In the next two chapters this difference in the application of the model will be seen to lead to strikingly divergent accounts of the nature of scientific progress.

---

[24] See e.g. Popper (1934), Ch. 3, where he treats theories in this way, explicitly stating on p. 59 that "Scientific theories are universal statements." In this regard see also Popper (1959), pp. 426ff.

CHAPTER 4

THE LOGICAL EMPIRICIST CONCEPTION OF SCIENTIFIC PROGRESS

1. A FORMAL CRITERION OF PROGRESS

As mentioned in Chapter 1, both Popper and the Empiricists advocate the use of the Deductive Model as a model of the explanation of particular occurrences. But the Empiricists go one step further and suggest its employment as a model of the explanation of laws by higher level theories. For example, Morris Cohen and Ernest Nagel have said: "Scientific explanation consists in subsuming under some rule or law which expresses an invariant character of a group of events, the particular event it is said to explain. Laws themselves may be explained, and in the same manner, by showing that they are consequences of more comprehensive theories."[1]

In presenting the basis of the Empiricist conception of science it was shown how the Deductive Model, originally introduced as a model of explanation, can also function as a model of induction and confirmation in the contexts of discovery and justification respectively. Here it may be put to the same use. Thus, on the Empiricist conception, scientific advance may be seen to consist in the discovery of higher level laws or theories which deductively entail lower level ones.[2] Following Cohen and Nagel this means that here the place of the explanandum in the Deductive Model is to be filled, not by an empirically verifiable statement, but rather by a universal law. This law L is to be derivable from some theory L_1 in conjunction with particular statements of conditions C_1, which can also have the

---

[1] Cohen & Nagel (1934), p. 397. But cf. Campbell (1921), p. 80: "To say that all gases expand when heated is not to explain why hydrogen expands when heated; it merely leads us to ask immediately why all gases expand. An explanation which leads immediately to another question of the same kind is no explanation at all."

[2] In his concentration on the work of Carnap, Lakatos misses this Empiricist conception of the growth of knowledge. See Lakatos (1968), p. 326.

character of universal statements. The inductive discovery and confirmative justification of L_1 is thus partly to depend on that of L. This may be schematized as below:

(10) $(L_1 \,\&\, C_1) \vdash L.$

To make explicit the deductive relation from L_1 and C_1 to L, (10) may be presented directly in the predicate calculus thus:

(11) $(y)(Hy \rightarrow Gy) \,\&\, (z)(Fz \rightarrow Hz) \vdash (x)(Fx \rightarrow Gx).$

And (10) and (11) may both be applied to an extension of the example used earlier:

(12)			
	L_1	$(y)(Hy \rightarrow Gy)$	All metals expand when heated
	C_1	$(z)(Fz \rightarrow Hz)$	Copper is a metal
	thus, L	$(x)(Fx \rightarrow Gx)$	All copper expands when heated.

If we conjoin to (10) the original schematization of the Deductive Model we get:

(13) $(L_1 \,\&\, C_1) \vdash L$ and $(L \,\&\, C) \vdash E.$

And (13) itself implies the original form of the model (as expressed in the sentential calculus):

(14) $L_1 \,\&\, (C \,\&\, C_1) \vdash E.$

In order that L_1 be more than a formal embellishment—that its discovery should constitute actual progress—a further requirement may be set. It is that L_1 have confirming instances, or make predictions, beyond those of L. This requirement can be easily handled in terms of the Deductive Model by conjoining to (13):

(15) $(L_1 \,\&\, C_2) \vdash L_2$ and $(L_2 \,\&\, C_3) \vdash E_1,$

where, as throughout, difference of subscript is to suggest difference of referent. And the conjunction of (13) and (15) may be exemplified by:

(16) L_1 All metals expand when heated

C_1	Copper is a metal	C_2	Tin is a metal
thus, L	All copper expands when heated	thus, L_2	All tin expands when heated
C	This is copper being heated	C_3	This is tin being heated
thus, E	This (copper) expands	thus, E_1	This (tin) expands.

The schematization of (16) in turn implies:

(17) $$L_1 \,\&\, (C \,\&\, C_1 \,\&\, C_2 \,\&\, C_3) \vdash E \,\&\, E_1,$$

which, except for there being more than one explanandum, is also of the form of the Deductive Model as expressed in the sentential calculus.

On the basis of the above an Empiricist criterion of scientific progress can be formulated in terms of the Deductive Model. Provided that all of the statements concerned are distinct and have been verified or confirmed, we should say that one law or theory L_1 is a progression beyond another, L, only if (a) L_1, in conjunction with certain statements of conditions C_1, entails L, and (b) L_1, in conjunction with statements of conditions C_2, entails other verified or confirmed statements which L alone in conjunction with C_2 does not entail, e.g., L_2.

Thus, in keeping with this criterion, we read in Hempel:

> [T]he uniformity expressed by Galileo's law for free fall can be explained by deduction from the general laws of mechanics and Newton's law of gravitation, in conjunction with statements specifying the mass and radius of the earth. Similarly, the uniformities expressed by the law of general optics can be explained by deductive subsumption under the principles of the wave theory of light.[3]

---

[3] Hempel (1962), pp. 100–101; see also Hempel (1965), pp. 343 ff. Though Hempel intends here to be providing examples of how the Deductive Model can be applied to the case of theories explaining laws, in a footnote he goes on to concede that: "strictly speaking, the theory [Newton's] contradicts Galileo's law, but shows the latter to hold true in very close approximation within a certain range of application. A similar relation obtains between the principles of wave optics and those of geometrical optics." (1962), p. 101 n. In regard to this point see the concluding paragraph of the present chapter.

2. THE PROBLEMS OF THEORETICAL TERMS AND CORRESPONDENCE RULES

As mentioned in Chapter 2, on the Positivist conception of meaningfulness a (non-tautological) statement is meaningful only if it is verifiable by direct experience. Thus, for the Positivist, meaningful claims should involve no 'descriptive' terms, i.e. no non-logical terms, the referents of which are not directly observable. While the Empiricist conception, on the other hand, replaces the verifiability requirement with that of confirmability, confirmable statements (empirical laws) should still be such as to admit no non-logical terms having unobservable referents. But when it is realized that scientific theories contain terms like "electron", the referents of which are not directly observable, the problem arises as to how statements containing such terms can be construed as meaningful, granting that the Empiricist programme wants to afford this status to all scientific terms. This is the problem of theoretical terms,[4] which also arises in the case of mensural terms like "temperature", and in the case of terms like "magnet", the referents of which have dispositional properties.

Where the Empiricist problem of theoretical terms concerns the transfer of meaning from the empirical level to the theoretical, their closely allied problem of correspondence rules may be said to start at the theoretical level, and to concern the question of how theoretical notions are to apply to empirical situations. On the present reconstruction the problem of correspondence rules may thus be seen to be that of showing how scientific theories, taken to contain theoretical terms as essential elements, can logically entail empirical laws, which are to contain no theoretical terms.[5]

One approach to the solution of problems of this sort has been to suggest means of eliminating theoretical terms in such a way as to retain the empirical consequences of the theory, as conceived on the

[4] Thus we see that it is the existence of theoretical terms which is the problem; the distinction between observational and theoretical terms was not made in order to *solve* the problem, as has been suggested e.g. in Putnam (1962) and Achinstein (1965).

[5] Cf. Carnap (1966*a*), p. 232: "The statement that empirical laws are derived from theoretical laws is an oversimplification. It is not possible to derive them directly because a theoretical law contains theoretical terms, whereas an empirical law contains only observable terms. This prevents any direct deduction of an empirical law from a theoretical one."

Empiricist view. For example, following a suggestion of F. P. Ramsey,[6] if we consider a scientific theory in conjunction with a statement of conditions to have the form:

(18) $(x)(Tx \rightarrow Ox)$ & Ta,

Where "T" is a theoretical term, "O" an observational term, and a some entity which it can be supposed has the property represented by the theoretical term, then "T" can be eliminated by reformulating (18) in terms of one of its consequences in the second-order predicate calculus:

(19) $(\exists\phi)((x)(\phi x \rightarrow Ox)$ & $\phi a)$.

Here the predicate constant T has been dropped in favour of a second-order (bound) predicate variable ϕ. And we see that (19) has in fact retained the empirical consequences of (18), namely, Oa.

However, while the 'Ramsey-sentence' (19) has succeeded in preserving the empirical consequences of (18) without employing the theoretical term "T", its doing so has required the introduction of the term "ϕ". And, as Hempel has said:

> [T]his means that the Ramsey-sentence associated with an interpreted theory T' avoids reference to hypothetical entities only in letter—replacing Latin constants by Greek variables—rather than in spirit. For it still asserts the existence of certain entities of the kind postulated by T', without guaranteeing any more than does T' that those entities are observables or at least fully characterizable in terms of observables. Hence, Ramsey-sentences provide no satisfactory way of avoiding theoretical concepts.[7]

But once again, more important than such questions as whether the 'Ramsey method' can be employed so as to overcome the Empiricists' problems of theoretical terms or correspondence rules is the fact that the 'theories' to which the method has been applied (by Ramsey himself, for example) bear little resemblance to actual scientific theories. And until such methods are shown to have application

---

[6] Cf. Ramsey (1931). We note however that Ramsey does not himself suggest the procedure given here as a means of eliminating theoretical terms, but as "[t]he best way to write our theory" (p. 231)—i.e. as the best way to make more salient those aspects of his example which he considers important.

[7] Hempel (1965), p. 216.

to real theories, even positive results would mean little with regard to our understanding of the nature of science.

3. THE PROBLEMS OF MEANING VARIANCE AND CONSISTENCY

If we construe the Empiricist notion of scientific progress as involving the relation between theories, rather than between a theory and a law, problems arise which have a more direct bearing on the topic of the present study. As has been pointed out by Paul Feyerabend, the conception of scientific advance as consisting in the deductive subsumption of earlier theories by later ones presupposes that the meanings of the terms common to the theories remain constant.[8] If this were not so, then such derivations (granting them possible) would at most tell us something about the syntactical relations between the theories viewed simply as parts of one and the same abstract system—they would tell us nothing about the semantical relations between them, i.e., about how the theories relate to each other as entities providing information about some aspect of the real world. Thus what will here be called 'the problem of meaning variance' is that in certain important cases of theory succession in actual science terms do undergo a change in their meanings (e.g. Newtonian "mass" vs. Einsteinian "mass"), and the conception of scientific theories as being related in ways suggested by the Deductive Model—which provides only the purported linguistic form of theories—is unable to take account of this fact.

A second point, stated clearly by Duhem,[9] and later emphasized by Popper,[10] and, following him, by Feyerabend,[11] is that in attempting to provide accounts of the same realm of phenomena it is often the case that successive theories (or laws) are inconsistent with one another. The above authors have expressed this by saying that such theories contradict each other. Later in this study it will be suggested that theory conflict does not take the form implied by the term "contradiction"; but here conceiving of science in the context of the

---

[8] See e.g. Feyerabend (1963), pp. 16 ff.
[9] Duhem (1906), p. 193.
[10] Popper (1949), pp. 357–359, and (1957), pp. 197 ff.
[11] See e.g. Feyerabend (1963), pp. 20 ff.

Deductive Model, we see that in the case where two successive theories do conflict, unless the succeeding theory (in conjunction with the relevant statements of conditions) is self-contradictory, it cannot formally entail its predecessor. This 'problem of consistency', i.e. this inability to provide an account of theory conflict, places severe limitations on the Empiricist conception of scientific progress. In the next chapter Popper's conception of scientific progress will be considered, and will be seen to employ the Deductive Model in such a way as to afford a conception of theory conflict.

CHAPTER 5

THE POPPERIAN CONCEPTION OF SCIENTIFIC PROGRESS

1. CONTRADICTION

In Chapter 3 the basis of Popper's philosophy of science, including his notions of basic statement, background knowledge, and corroboration, were shown to rest on the Deductive Model, in which scientific laws are conceived to be universal statements of the form: $(x)(Fx \rightarrow Gx)$. And in Chapter 4 an Empiricist conception of progress was presented, also in terms of the model. In the present chapter it will be shown that Popper's attempts to provide a conception of progress likewise rely on the Deductive Model, the basic difference between his view and that of the Empiricists being that, where the latter see succeeding theories as logically entailing their predecessors, Popper sees such theories as contradicting one another. Thus, for example, Popper claims that:

> [F]rom a logical point of view, Newton's theory, strictly speaking, contradicts both Galileo's and Kepler's .... For this reason it is impossible to derive Newton's theory from either Galileo's or Kepler's or both, whether by deduction or induction. For neither a deductive nor an inductive inference can ever proceed from consistent premises to a conclusion that formally contradicts the premises from which we started.[1]

In its simplest form, Popper's conception of the relation between successive theories can be schematized in terms of the Deductive Model by:

(20) $(L \& C) \vdash E$ and $(L_1 \& C) \vdash \neg E$,

where L and L_1 are the 'theories' in question, C are statements of conditions, E is an explanandum statement, and $\neg E$ its negation. (It may be noted that while on the Empiricist conception theories can be

[1] Popper (1957*a*), p. 198; see also Popper (1975), pp. 82–83 &n.

distinguished from laws in that they have the latter as their explananda, as mentioned in Chapter 3 on Popper's conception no distinction is drawn between laws and theories. In what follows the term "theory" will be used in the sense intended by Popper.)

The nature of Popper's conception of theory conflict can be further clarified by deriving from (20):

$$(21) \qquad (L \,\&\, C) \vdash \neg (L_1 \,\&\, C).$$

Here, however, it becomes clear that in order to obtain a contradiction between L and L_1 Popper requires the truth of statements C, just as he did in the case of his falsifiability criterion.[2] And so, strictly speaking, L and L_1 do not contradict one another. But in the present chapter this problem will be set aside, i.e. it will be allowed that C can in some way be determined true, and that Popper has thus succeeded in providing a notion of theory conflict (as contradiction) where the Empiricists have not.

Thus we see that where the essential relation between theories on the Empiricist view is that of deductive subsumption, on Popper's view it is formal contradiction. The Empiricists succeed in providing a conception of how one theory might be considered scientifically superior to another—but their view fails to account for theory conflict. Popper has provided a conception of theory conflict—now his is the task of showing how one of two such conflicting theories is to be conceived as constituting an advance over the other.

It might be thought that Popper should here simply say that, of two theories, the superior is the one which has not been falsified (or 'refuted'). But this will not do, for it does not treat of the interesting cases in which both theories are false. Were Popper to lay all his stress on falsification as a criterion for distinguishing good theories from bad, then all refuted theories would be equally unacceptable; and since on Popper's account the history of science shows that theories are constantly being falsified, he would have to say that science is not progressing at all, and thus commit himself to scepticism. Consequently, Popper must provide criteria allowing us to

[2] Formally, this is due to the conception of theories as hypothetical statements. Thus, for example, $(x)(Fx \rightarrow (Gx \,\&\, \neg Gx))$ is contradictory only if Fa is assumed true for some a.

determine, at least in principle, which of two contradicting false theories is a progression beyond the other.

In his attempt to provide such criteria there are two main factors that Popper must take into account. One has to do with how close each of the false theories is to being true. But this is not alone a sufficient basis for a criterion of progress, for it implies that some relatively trivial generalization might be a progression beyond some comprehensive, albeit refuted, theory. Consequently a second factor must be considered, namely the comprehensiveness or non-triviality of a theory. This latter factor constitutes Popper's concept of *content*, and a combination of both factors gives his notion of *verisimilitude*. Below, each of these concepts will be discussed in detail.

2. CONTENT

In Chapter 10 of *Conjectures and Refutations* Popper says:

> My study of the *content* of a theory (or of any statement [*sic*] whatsoever) was based on the simple and obvious idea that the informative content of the *conjunction*, *ab*, of any two statements, *a*, and *b*, will always be greater than, or at least equal to, that of any of its components.
>
> Let *a* be the statement 'It will rain on Friday'; *b* the statement 'It will be fine on Saturday'; and *ab* the statement 'It will rain on Friday and it will be fine on Saturday': it is then obvious that the informative content of this last statement, the conjunction *ab*, will exceed that of its component *a* and also that of its component *b*.[3]

On the Empiricist conception of progress the question of the relative content of theories is easily handled: the higher level theory has a greater content than the theory or law it subsumes. But Popper, in seeing theories as contradicting, requires some other means of determining relative content. It is peculiar then that the idea upon which his concept of content is based is formulated along Empiricist lines rather than along his own. The conjunction *ab* deductively subsumes each of *a* and *b* and thus has a content greater than or equal to either of them—but this is of no help in conceiving how the content of one proposition might be greater than that of another when the two of them contradict.

[3] Popper (1962), pp. 217–218.

Nevertheless, from this basis Popper goes on to explicate his notions of *probability of a theory*, and *testability*. He suggests that if *a* and *b* are empirical propositions, the probability that both are true is less than the probability that either alone is true. And since the falsification of either conjunct will falsify the whole, he says of *ab* that it is more testable than either *a* or *b* alone.[4] Popper considers this to provide a criterion of potential satisfactoriness of theories, and says:

> The thesis that the criterion here proposed actually dominates the progress of science can easily be illustrated with the help of historical examples. The theories of Kepler and Galileo were unified and superseded by Newton's logically stronger and better testable theory, and similarly Fresnel's and Faraday's by Maxwell's. Newton's theory, and Maxwell's, in their turn, were unified and superseded by Einstein's.[5]

Needless to say, this picture is not at all in keeping with Popper's view of successive theories as contradicting. It is more an Empiricist conception than a Popperian one, and its acceptance would immediately raise the question of how to account for theory conflict.

The problem of conceiving of increase in content in terms either of conjunction or deductive subsumption, while at the same time conceiving of the relation between competing theories as being that of contradiction, runs throughout Popper's writings on the topic of content. Though he cannot maintain both conceptions, and the latter is central to his philosophy of science, his explications of how one theory might have more content than another are almost always made along the lines of the former.

But in elaborating the concept of content Popper does provide a conception via which the sizes of the respective contents of contradicting theories might possibly be determined. This elaboration consists in a distinction between what Popper calls 'logical content' and 'empirical content': "the *logical content* of a statement or a theory *a* is the class of all statements which follow logically from *a*, while ... the *empirical content* of *a* [is] the class of all basic statements that contradict *a*."[6]

---

[4] Popper (1962), pp. 218 ff. Note that, as mentioned in Chapter 3, Popper's notion of testability differs from his notion of severity of test.

[5] Popper (1962), p. 220.

[6] Popper (1962), p. 232.

Popper's distinction between logical content and empirical content is essentially the distinction between the Empiricists' notion of a verifiable statement E derivable from a law L in conjunction with statements of conditions C, and Popper's own notion of a (falsifying) basic statement $(C \,\&\, \neg E)$ which contradicts L. The logical content of a theory is more its actual content—what the theory (plus statements of conditions) contains, and it is partly in terms of this notion that Popper later defines verisimilitude. At this point however, since there is no observational statement E that follows from L alone, Popper feels justified in claiming that the *empirical* content of L consists rather in the class of basic statements contradicting L.[7] But, as was seen above, Popper needs statements of conditions C for contradiction equally as much as the Empiricists need them for derivation, so the fact that E does not follow from L without C does not warrant granting the status of empirical content to the class of basic statements and not to the class of verifiable statements. (This conclusion is further supported by the fact that, seeing as on Popper's account competing theories contradict, at least part of the logical content of each must be in the empirical content of the other.)

While Popper denies the status of *empirical* content to the set of statements derivable from a theory, he goes on to argue that although the class of basic statements contradicting a theory is actually excluded by the theory, it may nevertheless be called the (empirical) *content* of the theory for the reason that its measure varies with that of the set of derivable statements:

> That the name 'empirical content' is justifiably applied to this class is seen from the fact that whenever the measures of the empirical contents, $ECt(t_1)$ and $ECt(t_2)$, of two *empirical* (i.e. non-metaphysical) theories, t_1 and t_2, are so related that
>
> (1) $$ECt(t_1) \leq ECt(t_2)$$
>
> holds, the measures of their logical contents will also be so related that
>
> (2) $$Ct(t_1) \leq Ct(t_2)$$
>
> will hold; and similar relations will hold for the equality of contents.[8]

---

[7] Popper (1962), p. 385.
[8] Ibid.

Thus through a measure of the respective classes of basic statements of two theories we should be able to determine the relative sizes of their (logical) contents. This process need not involve the deductive subsumption or conjunction of theories and so can allow for their contradicting; what it requires instead is some other means of determining which theory excludes the more basic statements.

But how, even in principle, is this determination to be made? Popper does not begin to answer this question, and in fact says that he sees no possibility of attaching numerical values other than 0 and 1 to measures of content.[9] Thus, at least as far as Popper is concerned, there is no measure of the classes of basic statements of each of two theories that can tell us which theory has the greater content.

Content as Varying Inversely to Absolute Probability

After presenting the concept of content in terms of conjunction (Chapter 10, *Conjectures and Refutations*), Popper goes on to say that content varies inversely to probability:

> Writing $Ct(a)$ for 'the content of the statement a', and $Ct(ab)$ for 'the content of the conjunction a and b', we have
>
> (1) $$Ct(a) \leqslant Ct(ab) \geqslant Ct(b)$$
>
> This contrasts with the corresponding law of the calculus of probability,
>
> (2) $$p(a) \geqslant p(ab) \leqslant p(b)$$
>
> where the inequality signs of (1) are inverted. Together these two laws, (1) and (2), state that with increasing content, probability decreases, and *vice versa*; or in other words, that content increases with increasing *im*probability.[10]

As mentioned above, Popper's line of thinking here is that the probability of the conjunction of two empirical propositions being true is less than the probability of either alone being true. Implicit in this is that the truth value of neither of the propositions is known—in other words, that the probability being considered is an a priori, or absolute, probability. Now, granting that content might vary inversely to absolute probability as Popper suggests, if a means of determin-

[9] Popper (1962), p. 397.
[10] Popper (1962), p. 218.

ing the respective absolute probabilities of two contradicting theories could be provided, then this would in turn afford a means of judging the relative sizes of their contents.

Essentially the same notion of absolute or a priori probability as considered by Popper is employed by a number of philosophers writing in the spirit of the logic of science, including Rudolf Carnap and William Kneale.[11] These philosophers would agree, following J. M. Keynes, that the a priori probability of each of the following formulae is greather than, or at least equal to, that of the one(s) above it:

(22) $(x)(Fx \rightarrow (Gx \ \& \ Hx))$
$(x)(Fx \rightarrow Gx)$
$(x)((Fx \ \& \ Ix) \rightarrow Gx).$

This is so because each formula is implied by those preceding it. Keynes warns, however, that we may not be able to compare the respective a priori probabilities of two generalizations unless the antecedent of the first is included in that of the second, and the consequent of the second is included in that of the first, as above.[12] This latter condition of course requires that, if the generalizations concerned are each consistent, then they do not contradict one another.

As a consequence of this, if Popper hopes to place any weight at all on the idea of one of two contradicting theories having a greater a priori probability than the other, then the onus is on him first to show how such a difference is to be conceived. But, as a matter of fact, he never even tries to show this. He instead takes the view that *all* laws or theories have absolute (and relative)[13] probability *zero*.[14] And so we must conclude that a measure of content is not to be found via consideration of Popper's notion of a priori probability.

---

[11] Cf. e.g. Carnap (1966 *b*), and Kneale (1964).
[12] Cf. Keynes (1921), p. 225.
[13] Cf. Popper (1959), p. 364. The notion of relative probability plays a small role in the present considerations. This is so because what is being demanded of Popper is a criterion for determining the relative superiority of *false* theories; and all such theories (on any account in which they are conceived as universal statements) must have relative or a posteriori probability zero.
[14] Cf. Popper (1958), p. 192, and (1959), pp. 363 ff. & 373.

Zero Probability and the Fine-Structure of Content

Popper's view that all theories have zero probability is arrived at in arguing against Empiricists such as Carnap, who have claimed that theories may have a higher or lower probability depending on their degree of confirmation. Popper's argument is in itself reasonable, and it is that any universal law or theory entails in infinite number of singular statements each with a probability less than one, and that since the product of the probabilities of these statements will equal zero, the probability of the law itself must equal zero.[15] But this leaves him in the awkward position of having to explain how the size of the content of theories might differ, considering that it varies inversely to probability.[16] In an attempt to rescue his notion of content Popper thus claims that, though the probabilities of any two universal statements must equal zero, an analysis of the 'fine-structure' of their contents may allow us to determine which content is the larger.

In order to exemplify how the content of two laws or theories can differ in spite of their both having zero probability, Popper suggests that we consider as laws: 'All planets move in circles' and 'All planets move in ellipses'.[17] He then goes on to say that, since the former entails the latter, it has the greater content and, furthermore, the greater degree of testability, since any test of the entailed law is also a test of the entailing one, but not vice versa.

Here again we see the employment of the Empiricist conception of succeeding theories deductively subsuming their predecessors, rather than contradicting them. Though Popper does not claim actual scientific theories to be related in exactly this way, he nevertheless suggests further that:

> Similar relationships may hold between two theories, a_1 and a_2, even if a_1 does not logically entail a_2, but entails instead a theory to which a_2 is a very good approximation. (Thus a_1 may be Newton's dynamics and a_2 may be Kepler's laws which do not follow from Newton's theory, but merely 'follow with good approximation'; . . . .)[18]

---

[15] Popper (1959), pp. 364 ff.

[16] Lakatos too notes that theories' having zero probability means that the notion of probability cannot be used to determine their content: see Lakatos (1968), p. 379.

[17] Popper (1959), p. 373.

[18] Popper (1959), p. 374.

Here we might expect Popper to conclude his argument by saying something to the effect that the actual case is sufficiently similar to that of deductive subsumption to warrant granting that a_1 has a greater content than a_2 (though Carnap could then employ the same argument to suggest that a_2 has a greater a priori probability than a_1). Or, considering his emphasis on the notion of testability in the present context, it might be thought that Popper should here say that even in the actual case a_1 is better testable than a_2, and consequently may be said to have the greater content. But what he does say is the exact converse—immediately following the above quotation Popper concludes that, "Here too, Newton's theory is better testable, because its content is greater."[19] But the question at issue does not directly concern the concept of testability—it concerns the concept of content. We see here, however, that Popper does not claim a_1 to have a greater content than a_2 on the basis of its being better testable, for, as he has himself suggested earlier, this should imply that any test of a_2 is also a test of a_1. And if he were to apply this reasoning to the present case, saying that any test of Kepler's laws is also a test of Newton's theory, he would be forced to admit that the falsification of Kepler's laws would entail the falsification of Newton's theory. This, of course, he cannot afford to do. Thus, aside from his not having provided a positive conception of difference of content in the case where successive theories are seen to contradict one another, Popper's attempt to overcome the problems raised by the zero probability of theories via the employment of the concept of logical entailment does not solve these problems, but only gives rise to further ones.

Nowhere in his dealings with the notion of content does Popper provide a conception of how to determine which of two contradicting theories has a greater content than the other. This is of major importance to his account of scientific progress, since according to it successive theories do contradict. What he does instead is present the notion of content in terms of the concepts of deductive subsumption and conjunction, which do not fit his view of science, and in terms of such concepts as those of absolute probability and testability, which are just as problematic as the notion of content itself.

---

[19] Ibid.

3. VERISIMILITUDE

The second major factor that must be taken into account in Popper's conception of scientific progress is that of 'nearness to the truth'. On Popper's view, though two contradicting theories may both have been refuted, one of them might nevertheless be closer to the truth than the other.

Two Conceptions of Nearness to the Truth

Taking theories to be entities that are either true or false, the idea of one false theory being closer to the truth than another may be thought of in one of two ways: either the theories may both be considered as being completely false, while one is closer to being true than the other; or they may be considered as each being partly true and partly false, while one is more true (or less false) than the other. In the former conception truth is not something attained, but rather something approached; in the latter conception truth, though attained, is not comprehensive.

Following Popper, these two conceptions may be exemplified by statements from ordinary language. An example of the first notion might go as follows: Given that the time at present is 12 o'clock, the statement 'The time is now 11 o'clock', while completely false, is closer to the truth than the also false 'The time is now 10 o'clock'.[20] And the second view might be exemplified by 'It always rains on Saturdays' and 'It always rains on Sundays'.[21] In this case Popper would consider both statements to be partly true in that it has rained at least once on each of these days. The former statement might be said to be nearer to the truth if it has rained more often on Saturdays than on Sundays.

The criterion for determining the distance from the truth of each of the completely false statements is an empirical one, involving the measurement of time (in this way it is similar to the criterion of scientific progress to be suggested later in this study); and it is difficult to see how it might be captured employing the tools afforded by the Deductive Model. On the other hand, the criterion of nearness

---

[20] Cf. Popper (1973), pp. 55 ff.
[21] Cf. Popper (1962), p. 233.

to the truth in the case where the statements are each conceived to be partly true appears, at least prima facie, to be susceptible of explication in terms of the model; and it is along this latter line that Popper develops his notion of verisimilitude.[22]

Verisimilitude and Logical Content

Popper's development of the notion of verisimilitude is in terms of the logical content of a theory being partly true and partly false. The class of non-tautological true statements in the logical content is the 'truth content', and the class of (all) false statements, the 'falsity content'.[23] In the present section, however, an earlier notion of verisimilitude given by Popper will be considered, in which the truth content of a theory is composed of the class of *all* true statements in its logical content.[24] This should simplify present considerations without weakening Popper's position.

Popper's conception of verisimilitude may be clearly presented in terms of the Deductive Model. (In what follows it will be assumed that all statements of conditions or background knowledge C_1 to C_4 are true, and laws L_1 and L_2 are false.) L_1, in conjunction with statements of conditions C_1, entails true explananda E_1; and in conjunction with C_2, gives false explananda E_2. Similarly, L_2 in conjunction with C_3 subsumes true explananda E_3, and in conjunction with C_4 gives false explananda E_4. In this way E_1 and E_3 are the respective truth contents of L_1 and L_2, and E_2 and E_4 are their respective falsity contents. These relations may be schematized:

(23)
$$(L_1 \ \& \ C_1) \vdash E_1,$$
$$(L_1 \ \& \ C_2) \vdash E_2,$$
$$(L_2 \ \& \ C_3) \vdash E_3,$$
$$(L_2 \ \& \ C_4) \vdash E_4.$$

---

[22] Although, as discussed in Chapter 3, Popper believes that no statement in empirical science can be determined to be true, he feels justified in speaking in terms of a theory as having certain true consequences thanks to the work of Alfred Tarski, to whom he ascribes a correspondence theory of truth: cf. e.g. Popper (1962), pp. 223 ff., and (1973), pp. 44 ff. For a critique of Popper's interpretation of Tarski, see Haack (1976); for problems concerning the applicability of Tarski's conception in the case of the comparison of theories, see Kuhn (1970 *c*), pp. 265 f.

[23] Popper (1973), pp. 47 ff.

[24] Cf. Popper (1962), pp. 391 ff.

At first it might be thought that a measure of verisimilitude for L_1 might be determined by subtracting the number of statements in its falsity content E_2 from the number in its truth content E_1, and that a measure for L_2 might similarly consist in subtracting the number of statements in E_4 from that in E_3.[25] But since each of the sets of explananda should contain an infinite number of statements, this method cannot be used. Popper thus suggests rather that two theories L_1 and L_2 be defined as comparable with regard to verisimilitude only in case: (a) the truth content of one of them, L_1, is a proper subset of the truth content of the other, L_2, while the falsity content of L_2 is identical with, or a proper subset of, the falsity content of L_1; or (b) the truth content of L_1 is identical with, or a proper subset of, the truth content of L_2, while the falsity content of L_2 is a proper subset of the falsity content of L_1.[26] In either case we should say that L_2 has a greater verisimilitude than L_1.

Following (23), Popper's definitions may be schematized in terms of the true and false explananda of each of L_1 and L_2:

(24) $(E_1 \subset E_3)$ and $(E_4 \subseteq E_2)$ implies that $(VL_1 < VL_2)$,

(25) $(E_1 \subseteq E_3)$ and $(E_4 \subset E_2)$ implies that $(VL_1 < VL_2)$.

Aside from the fact that Popper gives no examples of scientific theories which he believes actually to be related in this way, Pavel Tichý has proved that on Popper's definitions no false theory can have a comparably greater verisimilitude than any other.[27] In the case of (24), Tichý has shown that if the truth content of L_1 were a proper subset of the truth content of L_2, then it could not be the case that the falsity content of L_2 be identical with, or a proper subset of, the falsity content of L_1. This would be so for the following reasons: Since the truth content of L_1 is a proper subset of the truth content of L_2, there is an element in L_2's truth content that is not in that of L_1. And since L_2 is false, there is at least one element in its falsity content. The conjunction of these two elements is false, and since

[25] Cf. e.g. Popper (1962), p. 234.

[26] Cf. Popper (1972), p. 52. These definitions were deleted from Popper (1973), but were not replaced with alternatives affording a clear conception of how one of two false theories can have a greater verisimilitude than the other.

[27] See Tichý (1974).

each conjunct is in L_2, the conjunction itself is a part of L_2's falsity content. But it could not be a part of L_1's falsity content, for if it were, its true conjunct—just by being in L_1 and being true—would have to be in L_1's truth content, and this possibility has already been excluded. Thus it could not be the case that the falsity content of L_2 be identical with, or a proper subset of, the falsity content of L_1.

The above argument may be schematized:

(26) $(E_1 \subset E_3)$ and $(E_4 \subseteq E_2)$;
thus, $(\exists x)(x \in E_3 \;\&\; \neg(x \in E_1))$;
and, since L_2 is false,
$(\exists y)(y \in E_4)$;
thus, $(x \;\&\; y) \in E_4$;
but $\neg((x \;\&\; y) \in E_2)$, since $\neg(x \in E_1)$;
thus, $\neg(E_4 \subseteq E_2)$.

Tichý uses a similar argument to do away with (25) above. These arguments by Tichý, and similar ones by David Miller,[28] strike a serious blow to Popper's concept of verisimilitude. Though Popper might possibly get by these objections by having the notions of truth content and falsity content extend only to atomic statements and their negations, the efforts of Tichý and Miller have nevertheless succeeded in throwing a question mark on the concept of verisimilitude as a whole.

The considerations of the present chapter to this point have been concerned with the question of whether Popper has been able to avoid scepticism via the presentation of a conception of how one of two contradicting false theories may be thought to be a progression beyond the other, and to reveal how his attempts in this direction can be captured on the Deductive Model. It has been shown that his notion of content does not take account of theory conflict, and that his notion of verisimilitude is formally inadequate. We thus see that Popper has failed to afford a consistent conception of scientific progress that is in keeping with his view that successive theories contradict one another, and has consequently not succeeded in avoiding scepticism.

---

[28] See Miller (1974).

4–812581 Dilworth

The fact that Popper has not provided a consistent conception of how one false theory might be a progression beyond another does not mean that the provision of such a conception is impossible. But even if one were able to produce a clear account of theory succession which is in keeping with Popper's view, there exist problems of a more fundamental nature concerning the very applicability of the notions he employs in conceiving of science.

4. THE PROBLEMS OF MEANING VARIANCE AND THE NATURE OF SCIENTIFIC THEORIES

As presented in the previous chapter, the problem of meaning variance concerns the inability of accounts based on the Deductive Model to handle cases in actual science where terms such as "mass" undergo a change in meaning in their employment in different theories. Though Feyerabend has pointed to this problem in criticizing the Empiricist conception of theory succession, the fact that Popper's view is also based on the Deductive Model means that it arises here as well. Thus, just as saying that one theory entails another presupposes that their common terms have the same meaning, so does saying that one theory contradicts another. In other words, if there is a change in meaning of the terms common to two theories, then even if these theories were syntactically to contradict, this contradiction would be illusory. This in turn has relativistic consequences on the Popperian conception, for it implies that an observation statement purportedly corroborating one theory while falsifying another does not actually do so, since, given meaning variance, the terms it employs would not have the same meanings in both theories.[29] We see therefore that such a notion as Popper's concept of verisimilitude, even if it were successfully formulated in a consistent way, would fail to account for those cases in actual science where in the transition from one theory to another terms do change in meaning.[30]

The second problem to be mentioned here, that of the nature of

[29] This problem has come to be called 'the paradox of meaning variance'. For attempts at its resolution from within the Popperian framework, see e.g. Giedymin (1970) and Martin (1971).

[30] For similar comments see Kuhn (1970*c*), pp. 234–235.

scientific theories, is closely tied to the problem of meaning variance and to the Empiricists' problems of theoretical terms and correspondence rules. At this point this problem will be expressed merely as a doubt, based on the existence of these other problems, as to whether actual scientific theories do have the form of universal statements having truth values. In conjunction with this doubt, however, it may be pointed out that in actual science theories are often associated with *models* of one form or another, and neither Popper nor the Empiricists have shown how such scientific models might be related to theories conceived as universal statements.[31]

In the next chapter Imre Lakatos' development of Popper's views will be considered, as will Popper's 'three requirements for the growth of knowledge'. In both cases it will be found that to the extent that Popper and Lakatos rely on the Deductive Model they do not avoid the criticisms of the present section, and that to the extent that they do not, they no longer explain science, but merely describe it.

[31] Though Ernest Nagel, for example, has provided an interesting discussion of the nature of scientific theories and the possible role played by models in them, he has not made a direct attempt to explicate how models might function in such theories when conceived as universal statements: see Nagel (1961), Ch. 6.

CHAPTER 6

POPPER, LAKATOS, AND THE TRANSCENDENCE OF THE DEDUCTIVE MODEL

1. SOPHISTICATED METHODOLOGICAL FALSIFICATIONISM

One criticism of Popper's view that is suggested in the writings of Thomas Kuhn is that in actual science a theory is never rejected unless there is another theory to take its place.[1] Imre Lakatos recognizes a problem of this sort, but does not follow the path indicated by Kuhn, for he sees it as implying that theory change is an irrational process which can be analyzed solely from within the realm of (social) psychology.[2] The alternative Lakatos thus chooses is to develop further Popper's conception in the context of the Deductive Model[3] in an attempt to "escape Kuhn's strictures and present scientific revolutions not as constituting religious [*sic*] conversions but rather as rational progress."[4]

In the subsequent chapters of the present study however, it will be shown that the views of Kuhn (and Feyerabend) may be pursued in a natural way so as to eventuate in a conception in which the transition from one theory to its successor is based on both rational and empirical considerations. But in the present section Lakatos' conception —'sophisticated methodological falsificationism'—will be analyzed, and will be found to suffer from just that problem Lakatos ascribes to Kuhn, namely, the failure to provide rational grounds for preferring one of two competing theories to the other.

Lakatos' downfall in this regard is occasioned by his acceptance of

---

[1] Cf. Kuhn (1962), p. 77.

[2] Lakatos (1970), p. 93.

[3] That Lakatos is actually basing his reasoning on the Deductive Model should be clear not only from the fact that his views are readily reconstructible in terms of it, but also from his mention of the model e.g. in his (1970), p. 131, as well as his explicit use of it on pp. 185 ff. of the same article.

[4] Lakatos (1970), p. 93.

Popper's arguments that in the testing of a theory one can never be sure of having established true statements of initial conditions.[5] In agreeing with Kuhn that a theory tested in isolation is not to be considered invalid on the basis of the results of its tests, Lakatos presents 'an imaginary case of planetary misbehaviour' in which Newton's gravitational theory resists successive attempts at falsification for the reason that the conditions of each test are found to be different from what they were thought to be at the time of making the test.[6] Thus, according to Lakatos, since we can never be sure of the truth value of statements of initial conditions, we can never be sure that we have falsified a theory treated in isolation. But, as will be seen below, on the Deductive Model this reasoning applies not only to the present case, but also to that in which two theories are involved, and Lakatos does not succeed in showing falsification to be any more reasonable in the case of two theories than it is in the case of only one.

Lakatos presents his conception of the circumstances under which a theory may be considered falsified as follows:

> For the sophisticated falsificationist a scientific theory T is *falsified* if and only if another theory T' has been proposed with the following characteristics: (I) T' has excess empirical content over T: that is, it predicts *novel* facts, that is, facts improbable in the light of, or even forbidden, by T; (2) T' explains the previous success of T, that is, all the unrefuted content of T is included (within the limits of observational error) in the content of T'; and (3) some of the excess content of T' is corroborated.[7]

This conception is essentially the same as Lakatos' notion of one theory having a higher degree of corroboration than another,[8] and, setting aside his parenthetical qualification, both notions can be schematized in terms of the Deductive Model most simply thus:

---

[5] This view was discussed in Chapter 3; and the problems it faces were set aside in Chapter 5 in order to make more reasonable Popper's attempts to provide a conception of progress. It will here be taken up again since it plays a central role in Lakatos' considerations. Cf. Lakatos (1970), pp. 99 ff.

[6] Lakatos (1970), pp. 100–101. It may be suggested here however that Lakatos is missing Kuhn's point. Kuhn is not denying that a theory treated in isolation may be seen to have flaws—rather, he is suggesting that a theory is not *rejected* unless there is an alternative to take its place. See e.g. Kuhn (1962), p. 77.

[7] Lakatos (1970), p. 116.

[8] Cf. Lakatos (1968), pp. 375–384.

(27) $(L_1 \;\&\; C) \vdash E_t \;\&\; E_f;$
$(L_2 \;\&\; C) \vdash E_t \;\&\; \neg E_f.$

Here "L_1" is to denote theory T, which in conjunction with statements of conditions C implies certain true (or corroborated) empirical statements E_t, and certain false ones E_f. "L_2" denotes the superior theory T', which in conjunction with C entails the same true statements E_t, but instead of implying E_f implies $\neg E_f$.[9]

The above schematization makes it clear that, on Lakatos' reasoning, just as a theory treated in isolation cannot be falsified, neither can it be falsified in the presence of another theory. And this is so for the same reason, namely, that we can never be sure that our statements of conditions C are true. Thus, as is evidenced by (27), if we cannot be sure of the truth value of C, then we cannot know whether certain false statements are implied by one of the theories and not instead by the other. Though it might be argued that a more sophisticated schematization than (27) is required in order to capture Lakatos' reasoning here, any such representation would also require the determination of the truth of C, and would thus lead to the same result. Thus we see that Lakatos has himself failed to provide a positive conception in which one theory is preferable to another on the basis of rational considerations.

Lakatos in fact recognizes this problem, and devotes a good part of his paper "Falsification and the Methodology of Scientific Research Programmes" to wrestling with it in various forms. For example, in treating of the problem in the context of crucial experiments (in which, on the Deductive Model, C must be determined true), he in the end suggests that, "we may—*with long hindsight*—call an experiment crucial if it turns out to have provided a spectacular corroborating instance for the victorious programme and a failure for the defeated one".[10] But it is easy to see that this is no rational criterion of

---

[9] At one point in his writings Lakatos actually makes a suggestion implying that we should here treat statements of conditions C as being a *part* both of T and T': see Lakatos (1968), p. 382. While doing so would have no effect on the issues raised in the present discussion, it may nevertheless be pointed out that not only does this suggestion appear intuitively unsound, but Lakatos himself is not at all consistent in this regard. Cf. e.g. his above-mentioned 'imaginary case of planetary misbehaviour'.

[10] Lakatos (1970), p. 173.

acceptibility, for it does not tell us how long a hindsight we need; and even if it did draw the line at some point, that point would surely be arbitrary. As Feyerabend has said, "if you are permitted to wait, why not wait a little longer?"[11] Lacking an answer to this question, Lakatos succumbs to the very charges of irrationalism that he levels at Kuhn. But this is not all, for the question Lakatos ought really to be answering here is not when we should call an experiment crucial, but when we should consider a research programme victorious. And if he is to answer this question at all, his response in the present context must be to the effect that the victorious programme is the one which, with hindsight, is supported by crucial experiments. But, as we see, this reasoning only leads him in a circle.

However, a consideration of the above discussion might lead to the suggestion that to a large extent the critical points raised there do not directly affect Lakatos, but only show the Deductive Model to be unable to capture what he is saying. That the model does not fairly represent Lakatos is evident even in the case of his conception of falsification. There, for example, he speaks of part of the content of T being included, within the limits of observational error, in the content of T', and as has been admitted above, the Deductive Model does not capture this notion.

What it seems Lakatos is actually doing in such cases is working within the model, and then stepping outside it whenever convenient or necessary in order to make his position appear more reasonable. Considering that so much of what he says *is* reconstructible in terms of the model, and that he himself implies its presence in his considerations,[12] it would be naive to deny that the model is lying in the background throughout his works discussed here. This means, however, that these excursions into aspects of science which cannot be explained on the basis of the model are actually ad hoc, and while they are consequently almost invariably quite correct, they must be relegated to the status of mere description, and be seen to lie outside what can be accounted for on Lakatos' philosophy of science.

---

[11] Feyerabend (1970), p. 215; see also Kuhn (1970*c*), p. 262.

[12] Cf. footnote 3 of the present chapter.

2. POPPER'S 'THREE REQUIREMENTS FOR THE GROWTH OF KNOWLEDGE'

After discussing the notions of content and verisimilitude in Chapter 10 of *Conjectures and Refutations*, Popper goes on in the same chapter to suggest three requirements for the growth of knowledge. Popper's requirements, similar to a number of remarks made by Lakatos, seem for the most part to be quite reasonable, but they gain little support from his view of science based on the Deductive Model.

Popper's *first requirement* is that a new theory:

> should proceed from some *simple, new* and *powerful, unifying idea* about some connection or relation (such as gravitational attraction) between hitherto unconnected things (such as planets and apples) or facts (such as inertial and gravitational mass) or new 'theoretical entities' (such as field and particles).[13]

We might begin by investigating the extent to which Popper's philosophy of science supports our intuitive understanding of this requirement. That a new theory should proceed from some simple, powerful (logically strong) and unifying idea is, on Popper's view, all to say the same thing: that the theory have high content.[14] It may be pointed out, however, that Popper's notion of simplicity is itself thus counter-intuitive, since it suggests that the simplicity of a theory should increase with each empirical statement conjoined to it. Furthermore, Popper's requirement must be a relative one—i.e. the new idea must have a high content relative to that of the idea in the earlier theory. But on Popper's view the old and new theories are to contradict, and as has been shown in the previous chapter, Popper has not provided us with a concept of content that can be applied in such cases. Thus this aspect of his first requirement is either vacuous, or else it quite transcends what Popper is warranted in demanding given his basis in the Deductive Model.

A second point that may be raised concerns Popper's suggestion that a new theory should proceed from a *new* idea. On the Deductive Model, however, the potential viability of a theory should depend on

---

[13] Popper (1962), p. 241.

[14] Concerning 'simplicity' see e.g. Popper (1957 *b*), p. 61; for 'logical strength' see e.g. Popper (1962), p. 219; and for 'unification' see Popper (1962), p. 220.

its formal structure, either considered in isolation or in relation to the structure of other theories. Thus, unless it is intended tautologically, this suggestion is quite antithetical to Popper's position in the logic of science. And, in any case, viewed simply on its own merits, this demand not only seems unreasonable, but is at variance with the history of science (cf. e.g. the revitalization of the heliocentric conception, and the resurgence of atomism).

Thirdly, that Popper should speak of an *idea* as being distinct from the theory to which it leads is not at all indicated by his conception of theories as being universal statements. And lastly, Popper's reference to gravitational attraction, inertial and gravitational mass, and such theoretical entities as field and particles only gives rise to the problems of theoretical terms and correspondence rules, which are confounded in Popper's case by his having no criterion for distinguishing between laws and theories.

Thus while Popper's first requirement is of some merit, it nevertheless cannot be said to be warranted on the basis of his own philosophy of science in which scientific theories are universal statements of the sort suggested by the Deductive Model.

Popper's *second requirement* is that of independent testability:

> [A]part from explaining all the *explicanda* which the old theory was designed to explain, [the new theory] must have new and testable consequences (preferably consequences of a *new kind*); it must lead to the prediction of phenomena which have not so far been observed.[15]

Popper points out that this requirement is necessary (on his view) in order that the new theory not be ad hoc, for, since he makes no essential distinction between theories and statements, the new theory could otherwise just as well consist of a series of statements describing presently known phenomena.[16] His requirement that the new theory must have *new* consequences can be captured on the Deductive Model if we simply take it as meaning that it must have *different* consequences—but this would not prevent it from being ad hoc. On the other hand, Popper's suggestion that the new theory must lead to

---

[15] Popper (1962), p. 241.
[16] Popper (1962), pp. 241–242.

the prediction of phenomena which have not previously been observed, while it does imply that the theory not be ad hoc, is again quite beyond what is warranted on the basis of his position in the logic of science. On the Deductive Model what is of importance are formal relations and the truth values of such statements as those of initial conditions and explananda; *whether* certain of these truth values are known is of course of relevance, but, from a formal point of view, *when* they are known is not. Thus we see here that not only is Popper unjustified in making this second requirement his own, but his conception of science suffers yet another setback through its inability to distinguish ad hoc theories from those that are not.

Popper's *third requirement* is that some of the predictions demanded by the second requirement should be verified, and that the new theory should not be refuted too soon.[17] Here we see that, in presupposing the second requirement, and in suggesting that the new theory not be refuted too soon, Popper's third requirement also rests on temporal notions that find no place in his philosophy of science. But in elaborating this requirement Popper says that "the successful new predictions which we require the new theory to produce are identical with the crucial tests which it must pass in order . . . to be accepted as an advance upon its predecessor".[18]

This further clarification may be seen as suggesting a criterion free of temporal considerations; but to the extent that it is so viewed, Popper's third requirement also fails to distinguish between innovative and ad hoc theories. Furthermore, Popper's (and Lakatos') view that statements of conditions can never be determined to be true —while counter-indicated by the employment of the Deductive Model—gives rise to the problem discussed in the previous section, namely, that we can then never be sure that the results of a test actually favour one theory and not perhaps the other. And, setting this problem aside—i.e. assuming that the truth values of statements of conditions can be determined, Popper has still provided us with no consistent conception of how the superiority of one theory over another can be based on crucial tests in the case where both theories —conceived as universal statements—are false.

[17] Popper (1962), p. 247.
[18] Ibid.

Thus while Popper's third requirement, seen as suggesting that a superior theory pass certain crucial tests which its rival fails to pass, is quite reasonable and in keeping with a number of important episodes of the history of science, it at the same time lies beyond what Popper is justified in demanding from the basis of his own philosophy of science.

In reviewing the development of the present study to this point we see that both Popper's and the Empiricists' attempts to understand the nature of science and scientific progress through the employment of the Deductive Model face a good many difficulties. Perhaps the most important of these are not internal problems, but problems of application. As will be argued in the next chapter, it is largely the existence of such problems as concern the application of the Deductive Model that has led Kuhn and Feyerabend to claim that in certain cases of actual scientific advance the theories involved are incommensurable.

CHAPTER 7

KUHN, FEYERABEND, AND INCOMMENSURABILITY

1. 'INCOMMENSURABILITY' IN ITS NEGATIVE SENSE

The development of the Empiricist and Popperian conceptions of scientific progress in terms of the Deductive Model has shown each to suffer a serious drawback: the Empiricist view affords no conception of theory conflict, and the Popperian view provides no consistent conception of progress itself. Nevertheless, the intuitive notions motivating each of these philosophies of science, considered independently of the model, appear quite sound. One is still inclined to admit that, in some sense, succeeding theories do subsume their rivals, and that, in spite of this, such theories conflict with one another. Thus we might accept, for example, a description of theory succession in which the superior theory is said to explain both what its rival is able to explain, as well as certain of those states of affairs which are considered anomalous to the rival. But the problem here lies in the failure of the Deductive Model, and consequently the failure of both the Empiricist and Popperian conceptions of science, to provide an account of this sort of phenomenon. And, as regards the employment of the model itself, we may add to this the problem of meaning variance, and the very question as to whether scientific theories have the form of universal statements, as the model suggests.

This distinction between the provision of an adequate description of science on the one hand, and the provision of an explanation of science (based on the Deductive Model) on the other, might help us to understand the attitude of Thomas Kuhn toward the Popperian philosophy of science. While Kuhn agrees with many of the claims made by Popper and Lakatos, especially where they suggest a non-cumulative conception of theory succession, he nevertheless feels that there is a basic difference between his and their ways of seeing science. For example, as regards Popper, Kuhn has expressed the conviction that "our intentions are often quite different when we say the same things'';[1] and, after likening their differences to a switch of gestalt, Kuhn asks, ''How am I to show him what it would be like to wear my spectacles when he has already learned to look at everything I can point to through his own?''[2] In another place Kuhn describes his single most fundamental difference from Popper as resting in the latter's belief that the problem of theory choice can be resolved in such a way as is suggested by the notion of verisimilitude—a belief which shares with the views of Empiricists such as Carnap and Reichenbach the presupposition that ''canons of rationality [are to] derive exclusively from those of logical and linguistic syntax.''[3]

Now, it may be noted that Kuhn is not here, nor anywhere else, suggesting that science is not a rational enterprise. Quite the contrary: in keeping with a number of claims made by Popper and Lakatos he has in fact indicated more than once that he sees theory choice in science as being based on good reasons—reasons which are, furthermore, ''of exactly the kind standard in philosophy of science: accuracy, scope, simplicity, fruitfulness, and the like.''[4] But, from the point of view of the present study, where Kuhn does differ from Popper and Lakatos may be said to be in his view that the nature of theory succession in science, and with it the very nature of scientific rationality itself, does not have the form suggested by the employment of the Deductive Model.

This way of seeing the main difference between Kuhn's view and those of the Empiricists and Popperians suggests that such charges of irrationalism as have been levelled at Kuhn by, for example, Lakatos presuppose the very basis Kuhn is calling into question. What seems to be the case is that many of Kuhn's critics presuppose, either wittingly or otherwise, that an account of science as a rational enterprise must take the form suggested by either the Deductive Model or

---

[1] Kuhn (1970*b*), p. 3.

[2] Ibid. As regards Lakatos, see Kuhn (1971).

[3] Kuhn (1970*c*), p. 234. See also pp. 266–267 of the same article, and Kuhn (1974), p. 504: ''My main and most persistent criticism of the recent tradition in philosophy of science has been its total restriction of attention to syntactic at the expense of semantic problems.''

[4] Kuhn (1970*c*), p. 261. See also, e.g., his (1970*a*), p. 199. For an instance of the belief that Kuhn sees the decision between scientific theories as being irrational, see Laudan (1977), pp. 3f. & 141.

some other formal construction, and furthermore, that any alternative approach automatically lies outside what can correctly be called the philosophy of science.[5] But this view has nowhere been supported —let alone established—by argument; and, considering that the Deductive Model suffers so many failings, one wonders if it would not be wiser to accept Kuhn's view as an invitation to the development of an alternative, non-formal, conception of science as a rational (and empirical) enterprise.

The dissatisfaction expressed by Kuhn with regard to the Empiricist and Popperian conceptions of science and scientific progress is largely shared by Paul Feyerabend. But where Kuhn approaches science mainly from the point of view of an historian, Feyerabend, while he does argue from historical examples, takes more the approach of a philosopher of science. Having early studied under Popper, Feyerabend was quick to see the failings of the Empiricist view, pointing out its problem of meaning variance as well as emphasizing its problem of consistency, as has been mentioned earlier in this study. And in his somewhat more recent work Feyerabend has also come to criticize the Popperian notion of verisimilitude for reasons pertaining to meaning variance, viz., that because of the differing concepts behind their common terms the respective contents of certain theories cannot be compared.[6]

Thus, when it comes to the question of the nature of theory succession in science, the present study leads to the view that the essence of both Kuhn's and Feyerabend's dissatisfaction with the Empiricist and Popperian accounts may be seen as stemming from problems faced in the attempted application of the Deductive Model to actual science. And, to take this suggestion one step further, we may say that the respective claims of Kuhn and Feyerabend that in certain cases of theory succession the theories concerned are *incommensurable* amount to, *in their negative sense*, the claim that the Deductive Model (or, as might also be urged, any formal construction) fails to provide a realistic conception of the phenomenon as it actually occurs in science.

---

[5] This view is still held by influential commentators today: see e.g. Stegmüller (1979), p. 69.

[6] Cf. e.g. Feyerabend (1970), pp. 220–222.

2. 'INCOMMENSURABILITY' IN ITS POSITIVE SENSE

The term "incommensurable" is defined in the *Concise Oxford Dictionary* (1934 & 1977) as: "(Of magnitudes) having no common measure integral or fractional (*with* another); irrational, surd; not comparable in respect of magnitude; not worthy to be measured *with*". Kuhn and Feyerabend each borrow the term from this mathematical or mensural context and use it in attempts to express what is perhaps the same particular insight concerning the nature of the relation between certain scientific theories. It may be noted however that they do not apply the term to exactly the same cases. While both agree that such theories as Newtonian and Einsteinian mechanics are incommensurable,[7] where Kuhn would consider, for example, the geocentric and heliocentric systems of astronomy also to be so, Feyerabend would not.[8] Certain differences in their respective applications of the notion will be noted as this study proceeds, but at this point emphasis will be placed on what is common to their views, in an attempt to delineate one central positive aspect of 'incommensurability' capable of providing a starting point for the development of an alternative conception of theory succession in science.

On almost every occasion in which Kuhn speaks of incommensurability, and at a number of places where Feyerabend does so, those points of view which are taken to be incommensurable are at the same time also seen as being competing or incompatible.[9] Keeping this in mind we may also note that, as is suggested by Feyerabend's critique of Empiricism, an important rationale for the claim that certain theories are incommensurable is the suggestion that there is a change in the meaning of some or all of the 'descriptive' terms common to the theories in question when one moves from one of them to another. For example, in considering the transition from classical celestial mechanics to general relativity Feyerabend has said, "the meanings of all descriptive terms of the two theories, primitive as well as

[7] Cf. e.g. Kuhn (1962), p. 102, and Feyerabend (1965*b*), pp. 230ff.

[8] For Kuhn, cf. his (1962), p. 102; Feyerabend does not include this case among his examples in, e.g., his (1975), pp. 276–277, and has pointed out in a personal communication that he does not take it to constitute an instance of incommensurability.

[9] Cf. e.g. Kuhn (1962), pp. 103 & 165, and (1970*a*), p. 175; and see Feyerabend (1975), p. 274.

defined terms, will be different'',[10] and, speaking more generally, Kuhn has said, ''In the transition from one theory to the next words change their meanings or conditions of applicability in subtle ways''.[11]

As has been noted in earlier chapters, the viability of such claims concerning meaning change creates important problems as regards the applicability of the Deductive Model, and leads directly to relativistic consequences in the case where theories are seen to contradict. Unfortunately, however, it seems that those who have dealt with this problem have assumed that the Deductive Model or some similar basis employing the first-order predicate calculus is the only possible vehicle for the analysis of theory change in science. As a result of this, the relativism that arises in such contexts has been taken to imply a relativism in Kuhn's and Feyerabend's claims themselves; and it has not been realized that in this particular case, granting that such changes in meaning do occur, it is rather the Popperian conception that leads to relativism, as has been suggested in Chapter 5.

Though both Kuhn and Feyerabend have in their earlier writings based their claims that certain theories are incommensurable largely on this change of meaning of individual terms, neither of them has identified incommensurability with meaning variance. In fact, in reaction to his critics' coming more or less to equate these two notions, Feyerabend has said such things as that ''in the decision between competing theories, ''meanings'' play a negligible part'',[12] and that ''As far as I am concerned even the most detailed conversations about meanings belong in the gossip columns and have no place in the theory of knowledge.''[13] Both Kuhn and Feyerabend have emphasized rather that the shift between incommensurable theories involves a *fundamental* change, and in a recent work Feyerabend has instead begun his discussion of incommensurability by taking it to be exemplified by gestalt switch phenomena.[14] In this same vein, Kuhn in his

---

[10] Feyerabend (1965 *b*), p. 231.

[11] Kuhn (1970 *c*), p. 266. Note that Kuhn does not claim that *all* descriptive terms should change in meaning. In this regard cf. also p. 506 of (the discussion following) Kuhn (1974).

[12] Feyerabend (1965 *a*), p. 267.

[13] Feyerabend (1965 *b*), p. 230.

[14] Feyerabend (1975), pp. 225 ff.

earlier writings has suggested that "the switch of gestalt, particularly because it is today so familiar, is a useful elementary prototype for what occurs in full-scale paradigm shift",[15] and furthermore that "Just because it is a transition between incommensurables, the transition between competing paradigms cannot be made a step at a time, forced by logic and neutral experience. Like the gestalt switch, it must occur all at once (though not necessarily in an instant) or not at all."[16]

Following this lead, the essence of the notion of *incommensurability—in its positive sense*—will here be taken to be the idea of being related as are the different aspects of a gestalt switch figure. In what follows of this study this idea will be developed so as to provide a general characterization of theory change which is applicable to any case of competing scientific theories, and which not only avoids the problem of relativism that meaning change gives rise to on the Deductive Model, but is capable of explaining both theory conflict and scientific progress.

[15] Kuhn (1962), p. 85.
[16] Kuhn (1962), p. 150.

CHAPTER 8

THE GESTALT MODEL

1. A MODEL OF THEORY CHANGE VS. AN EXAMPLE OF PERCEPTUAL CHANGE

Gestalt switch diagrams have perhaps most often been used as paradigmatic examples of entities which can be perceived in completely different ways without their changing, and without there being a change in the perceiver's physical relation to them. N. R. Hanson, pursuing Wittgenstein's remarks concerning seeing and 'seeing as',[1] has employed gestalt switch figures in this sort of way in considering cases relevant to the philosophy of science. In the chapter of his book *Patterns of Discovery* entitled "Observations", for example, he has followed this line of thought, suggesting that where "Tycho and Simplicius see a mobile sun, Kepler and Galileo see a static sun."[2]

Kuhn, though he is aware of the difficulties involved in the idea that those employing different theories need *see* the world differently,[3] nevertheless devotes Chapter X of *The Structure of Scientific Revolutions* to a development of his own thoughts along this line. Feyerabend too is attracted to the idea, but feels he cannot accept it: "I strongly sympathize with [Hanson's] view, but I must now regretfully admit that it is incorrect. Experiments have shown that not every belief leaves its trace in the perceptual world and that some fundamental ideas may be held without any effect upon perception."[4]

But if we follow Hanson further, we see that in the chapter of his book entitled "Theories" he employs the gestalt switch phenomenon not so much as an example at the level of observation, but as an

---

[1] Cf. Wittgenstein (1953), IIxi.

[2] Hanson (1958), p. 17. It may be noted that this is perhaps the most extreme thing that Hanson says in this regard, and that he devotes more space to a discussion of the differences between a layman's and a scientist's ways of seeing a piece of scientific apparatus.

[3] Cf. Kuhn (1962), pp. 88 & 113ff.

[4] Feyerabend (1965*b*), p. 247. See also Feyerabend (1977), p. 365n.

analogue to the level of theory. Here, in the context of criticizing accounts based on the Deductive Model, he argues by analogy from a gestalt switch figure that there is an important distinction between observational detail statements and the conceptual frameworks (theories) within which they are cast.[5]

Jan Andersson and Mats Furberg, employing the same gestalt switch figure as Hanson, have obtained similar sorts of results which, though intended to provide insight regarding the nature of perception, may be seen from the point of view of the present study to provide the outline of a *model* of how different theories relate one to another.[6]

2. THE DUCK-RABBIT AS A MODEL OF THEORY SUCCESSION IN SCIENCE

It is intended that the term "model" as used here have the meaning it usually has when employed in everyday and scientific contexts—a meaning which is closely related to the notion of analogy. Whether what follows be called a model or an analogy is thus not of great concern to the present study, but seeing as the gestalt switch figure to be treated below is being advanced especially to serve as the *basis* of the account of the next chapter, and is not being introduced after the fact in order to support certain singular conclusions, it seems more in keeping with ordinary usage to call it a model.

The following presentation of the Gestalt Model will thus consist in the pointing out of certain features of a gestalt switch figure—features

[5] Cf. Hanson (1958), p. 87: "Consider the bird-antelope in fig. 12. Now it has additional lines. Were this flashed on to a screen I might say 'It has four feathers'. I may be wrong: that the number of wiggly lines on the figure is other than four is a conceptual possibility. 'It has four feathers' is thus falsifiable, empirical. It is an observation statement. To determine its truth we need only put the figure on the screen again and count the lines.

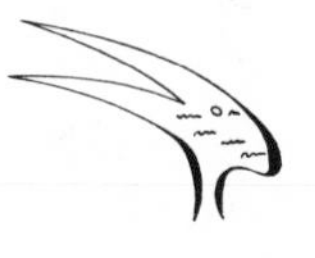

The statement that the figure is of a bird, however, is not falsifiable in the same sense. Its negation does not represent the same conceptual possibility, for it concerns not an observational detail but the very pattern which makes those details intelligible. . . . "

[6] Andersson's and Furberg's treatment of the gestalt switch phenomenon thus differs from the one to follow in the present chapter in that they do not consciously present it as a basis for an alternative conception of science: cf. Andersson & Furberg (1966), Chs. 5–7.

which in the next chapter will be seen to provide a foundation for the explanation of many important aspects of the relation between actual scientific theories. In contradistinction to the use of such a figure in the context of discussing certain perceptual phenomena, emphasis will here be placed on particular *conceptual* features that it exhibits.

What follows of this chapter thus constitutes only a presentation of the model, and not its employment. But it may be kept in mind that the remarks to be made here are intended ultimately to have relevance to full-fledged theories—theories such as Galileo's terrestrial physics and Newton's theory of gravitation.

The Gestalt Model differs from the Deductive Model in regard to flexibility. *Any* gestalt switch phenomenon may be used as a basis for considering the nature of the relation between particular theories. One may use, for example, Hanson's bird-antelope, or Wittgenstein's duck-rabbit[7]—whatever gestalt switch figure best suits the peculiarities of the theories to be considered. In this study the duck-rabbit will be taken as the model, since it is capable of affording a starting point for a relatively general account of the nature of the relation between successive theories.

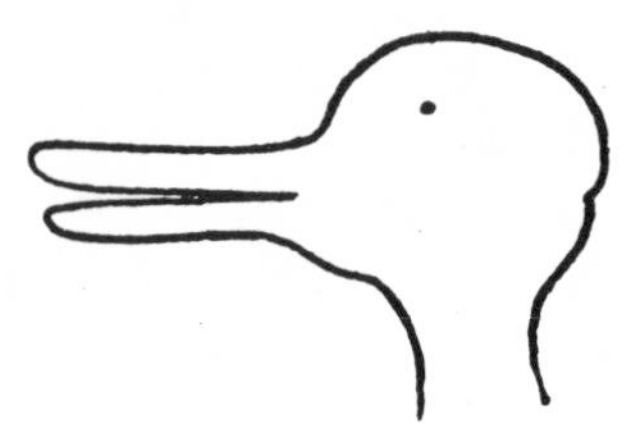

Figure.

We might begin by asking what it is that is peculiar about the above figure so as to justify our calling it a gestalt switch figure, and thus take it as differing in a special way from more ordinary drawings, such as Wittgenstein's 'picture-face'.[8] While it may be difficult to describe the difference in a way which everyone would find acceptable, we might try by suggesting that where the picture-face can be

---

[7] Wittgenstein (1953), p. 194.
[8] Ibid.

seen only in one way, the duck-rabbit can be seen in two.[9] And to this we might add that, as a matter of psychological fact, the duck-rabbit cannot be seen both as a picture-duck and a picture-rabbit by one person at one and the same time, in the sense in which, say, a picture of a red ball can be seen at one time to be of an object which is both red and round.[10]

3. THE SEEING OF AN ASPECT AS THE APPLICATION OF A CONCEPT

The shift from seeing the duck-rabbit as a picture-duck to seeing it as a picture-rabbit, or vice versa, Wittgenstein calls a 'change of aspect'. Each aspect is quite general in the sense that it exhausts the whole of the figure, and yet each is also exclusive of the other. In what follows the seeing of the duck-rabbit as a (picture-) duck[11] (or seeing it under the duck aspect) will be taken to be the application of the concept 'duck' to it, and thus here we may say that the shift from applying the concept 'duck' to applying the concept 'rabbit' is a fundamental and revolutionary one—the figure is seen completely differently once the other concept is applied.

If, upon first looking at the duck-rabbit, a person were able only to see the duck aspect, he could be helped to see the rabbit aspect by being told that what for the duck is its bill, for the rabbit is its ears, and so on. We would not argue or try to prove that the figure can be seen as a rabbit; instead we would try to *show* the person the rabbit aspect. A second way of doing so, rather than 'translate' from the duck aspect as above, would be to translate from a more neutral

[9] This has sometimes been misleadingly expressed by saying that gestalt figures are ambiguous. But where we might say of some linguistic entity such as a phrase or sentence that it is ambiguous, it would at least be metaphorical to say the same of gestalt figures.

[10] As expressed by Kuhn: "Though most people can readily see the duck and the rabbit alternately, no amount of ocular exercise and strain will educe a duck-rabbit." (1977) p. 6.

Note also that a treatment of the question as to *why* the duck-rabbit has the particular features it does is perhaps best handled by a psychologist; at any rate, such a treatment lies beyond the scope of the present study.

[11] Following Wittgenstein, in this study such terms as "picture-duck" will often be dropped in favour of the simpler formulation "duck". In all cases however these terms are meant to have reference only to the duck-rabbit figure.

description of the figure itself, telling the person for example that the long and narrow part of the figure is for the rabbit its ears, and that the indentation on the right side is its its mouth, etc. Or a third way might be to use ostension, and simply point out parts of the rabbit aspect, naming them as we do so.

Once the person has seen both aspects he might easily compare them; for example, he could say that the duck is looking to the left, while the rabbit is looking up, or to the right.[12]

4. SIMULTANEOUS APPLICATION

As mentioned above, as a matter of fact the duck-rabbit cannot be seen both as a duck and as a rabbit at the same time by the same person; or, in other words, one person cannot *simultaneously* apply the concept 'duck' and the concept 'rabbit' to the duck-rabbit.[13] However, the two concepts can of course be applied simultaneously if each is being applied by a different person; and, as will be treated below, this is also the case if each concept is being applied to a different figure.

5. ASPECTUAL INCOMPATIBILITY

Perhaps the feature of the Gestalt Model which is the most difficult to describe is also that which makes gestalt switch figures themselves so interesting, namely, the *conflict* that ensues when one attempts to see both aspects of such a figure at the same time. What is the nature of this sort of conflict or incompatibility, and can it be described in terms with which we are familiar independently of the existence of gestalt figures themselves?

If we compare the Gestalt Model and the Deductive Model on this point, we see that the sort of incompatibility that the Deductive Model can be used to express is essentially that of one sentence or

[12] Cf. Andersson & Furberg (1966), p. 49: "the antelope head looks to the right, the pelican head to the left; the antelope head but not the pelican head has clearly distinguishable horns; etc."

[13] Similarly, Wittgenstein says, though with different emphasis: "But the *impression* is not simultaneously of a picture-duck a picture-rabbit." (1953), p. 199.

statement negating another; in other words, conflict is there depicted as being of a linguistic nature, and as basically consisting in one law contradicting or denying what is entailed by another (plus statements of conditions). But in the present case we are not dealing with the making of statements or assertions so much as with the application of concepts. Upon seeing the duck-rabbit as a rabbit, one does not say, even to oneself, 'this is a rabbit'; rather, one simply applies the concept 'rabbit' to the figure. The sort of conflict that can arise here is prior to the employment of language; i.e., we might say that the conflict, while involving concepts, is pre-linguistic, for one's inability to see the duck-rabbit as both a duck and a rabbit at the same time is quite independent of his having words such as "duck" and "rabbit" to employ in describing the two aspects.

However, by continuing to speak in terms of the application of concepts we can come somewhat closer to a neutral description of the sort of conflict involved here. To this end then we should say that the concepts 'duck' and 'rabbit' are incompatible when the attempt is made to apply them simultaneously to the same thing. But even working at the more abstract level of concepts, we find that the situation cannot correctly be described as being essentially the same as that involving contradiction. This is so because a *contradiction* consists in the simultaneous affirmation and denial of one and the same proposition,[14] and not only are there no propositions or statements involved in the present case, but neither is there a notion of denial or negation. Rather, what we have here is more similar to what Aristotle would consider to be *contrary qualities* where, for example, he says, "nothing admits contrary qualities at one and the same moment".[15] But, in order to avoid misunderstanding, the sort of conflict considered here will simply be termed 'aspectual incompati-

---

[14] This notion of contradiction should be that which philosophers generally intend when they employ the term. Cf. e.g. Aristotle *De Int.* 17^a 31–33: "every affirmation has an opposite denial, and similarly every denial an opposite affirmation. We will call such a pair of propositions a pair of contradictories."

[15] *Cat.* 5^b 39–6^a 1; cf. also *Cat.* 14^a 11–13, and *De Int.* 24^b 8–9: "contrary conditions cannot subsist at one and the same time in the same subject." Note however that for Aristotle contrary qualities are exemplified by such entities as the colours black and white, which differ most widely within the same 'class' or sub-category. Cf. e.g. *De Int.* 23^b 22–24.

bility', and will be understood as involving the simultaneous application of differing concepts to one and the same thing, but not as involving the use of propositions, nor, consequently, as being the result of one proposition or statement negating or denying what is asserted by another.[16]

6. THE UNIQUENESS OF THE REFERENT

In order for conflict to arise in the case of the duck-rabbit not only is it necessary that the concepts 'duck' and 'rabbit' be applied at the same time, but, as mentioned earlier, both must be applied to the same figure or thing. Thus, for example, if we had two duck-rabbit figures side by side we might simultaneously see one of them as a duck and the other as a rabbit. In such a case there would be no conflict. Expressing this in terms of concepts we might thus say that conflict arises in the present case only if it is the *intention* of the person employing the concepts that both be applied to the same thing, and that when this is so both may be said to have the same *referent*. In other words, we can say that the concepts involved are not incompatible in and of themselves, but that they conflict only when they are given the same *reference*.[17]

7. PREDICATES OF THE SAME CATEGORY

The fact that two differing concepts or predicates be simultaneously given the same reference is however not itself sufficient for conflict to arise, since we might see one and the same thing as being, for example, both red and round; i.e., we can apply the predicates 'red' and 'round' to the same thing without conflict ensuing. On the other hand, we cannot see (the whole of) one thing as simultaneously being both red and blue. The difference between these two cases will here be expressed by saying that in the latter case the predicates 'red' and

---

[16] The difficulties involved in conceiving of the present sort of case in terms of contradiction will be further explored in the next chapter.

[17] Note that this usage of the terms "referent" and "reference" is independent of any linguistic connotation they might normally have.

'blue' are of the *same category* 'colour',[18] whereas in the former case the predicates concerned were respectively of the categories 'colour' and 'shape' or 'form'.

Employing this line of thought in the case of the duck-rabbit we should thus say that the concepts or predicates 'picture-duck' and 'picture-rabbit' are of the same category 'shape', and that their being of the same category is a prerequisite for their conflicting when simultaneously applied to the duck-rabbit.

8. RELATIVE ACCEPTABILITY: ACCURACY, SCOPE, AND SIMPLICITY

Once a person has been able to see the duck-rabbit both as a duck and as a rabbit, the question might arise as to which concept is the better applicable.[19] However, assuming that this person already has a definite view on the matter and considers e.g. the rabbit concept for some reason to be superior in this regard, if he were to attempt to convince another person who has only been able to see the duck aspect that this is so, the ensuing debate might well be characterized as a 'talking at cross purposes'. For example, what the first person calls the rabbit's ears, the second sees only as the duck's bill, and he might thus consider all the other's talk about 'ears' as being utter nonsense.

Accuracy

On the other hand, if the two parties are each able to see both aspects of the duck-rabbit, a meaningful argument might follow regarding whether the figure is more like a duck than a rabbit or vice versa. On first consideration it might appear that there is nothing in the figure to favour either aspect; but a closer inspection reveals that the indentation which is the rabbit's mouth does not at all seem in keeping with the concept of the duck-rabbit as a duck.[20]

---

[18] In this study categories will be treated as abstract entities, i.e. as entities which are of the sort to which predicates or concepts, rather than properties, belong; thus we will speak of the category: 'colour', rather than of the category: colour.

[19] As expressed by Andersson and Furberg: "Which way of seeing fig. 3 is the more correct, that of the person seeing it as an antelope, or that of the one seeing it as a pelican?" (1966), p. 54.

[20] Cf. also Andersson & Furberg (1966), p. 55: "The correctness of seeing a phenomenon under a certain aspect is supported then, if new perceptions of other parts of the

Assuming that we are unable to find anything in the figure which can be accounted for in the duck aspect while being anomalous to the rabbit aspect, we might then be inclined to say that the rabbit concept is the more *accurate*. Any part of the figure which may be criterial for the application of the concept 'duck' to the duck-rabbit is also criterial for the application of the concept 'rabbit'; but there is 'evidence' that favours the application of the rabbit concept that is also evidence against the application of the duck concept. Thus, though the two concepts can conflict prior to the discovery of something anomalous to one but not to the other, it is only with such a discovery that they can be judged as to their relative accuracy.

The indentation which is crucial to the determination of the relative accuracy of the two concepts might not have been visible to the naked eye. In such a case, the person advocating that the figure is more rabbit-like than duck-like might have suggested trying to find the notch (the rabbit's mouth) by using an instrument such as a magnifying glass. The unavailability of such an instrument would, in this case, have precluded the determination of either concept being more accurate than the other.

Scope

We might also imagine for the moment a slightly different case in which the figure contains not only the duck-rabbit head, but also the outline of a whole body. And let us say that one of the two concepts is clearly better applicable to the whole of this larger figure. (Note that the duck-rabbit is not to be thought of as actually *being* a figure of a duck as versus that of a rabbit, or vice versa. For present purposes the figure might just as well have been created by chance; and the questions being raised here simply concern the relative applicability of two particular concepts to it.) In this case then we should say that the concept which is the better applicable to the larger figure has a

---

phenomenon (or of the same parts but in new ways) strengthens the phenomenon's falling under the concept in question.

When the seeing of an aspect allows a perceptual filling out in this way, its correctness can be tested in about the same way as can that of a scientific theory." Note however that here reference is being made to scientific theory in order to help clarify the nature of seeing under an aspect, and not vice versa, as in the present study.

greater *scope* than does the other, or, we might say that it is the more *general* concept of the two.

Simplicity

But even granting the existence of the indentation, the supporter of the duck concept might have rejoined by saying that the duck-rabbit nevertheless closely resembles a duck—one which has suffered a blow to the back of its head! But if this were all he had to say, his claim would have to be recognized as being merely *ad hoc*, for, where the notch is integral to the rabbit concept (as being the rabbit's mouth), in the duck concept it is, in a sense, something 'added on'. Thus while the concepts 'rabbit' and 'duck that has been struck in the back of the head' might be equally accurate in accounting for why the figure appears as it does, other things being equal the former concept should be said to be the *simpler* of the two.

Should this mean then the abandonment of the duck concept? Not at all. Though the rabbit concept may be the more accurate, it might be pointed out that, for example, on first glance the figure still *looks* more like a duck.

Similarly, the discovery of the anomalous notch *prior* to the seeing of the duck-rabbit as a rabbit should not have led to the rejection of the duck concept (though it might well have led to one's wondering whether there exists another concept in which the notch plays an integral role). To conceive of the figure as a duck with an unexplained indentation in the back of its head is better than to have no clear concept of it at all.

Further scrutiny of the duck-rabbit reveals a bump behind the rabbit's ears (under the duck's bill) which we might here consider anomalous to both aspects. Again though, the presence of an aberration should not lead to the elimination of either concept.

The duck-rabbit is a figure which may be seen either as a picture-duck or as a picture-rabbit, and is not simply a picture of a duck and not that of a rabbit, or vice versa. Neither of the aspects in which the figure may be conceived can be determined in any absolute sense to be better than the other, and neither should be discarded. Though there is a sense in which the rabbit aspect both is superior to and subsumes the duck aspect, this superiority does not rest in the latter's being formally deducible from, or reducible to, the former; and though there is a sense in which the duck aspect is incompatible with the rabbit aspect, this incompatibility is not essentially one of contradiction.

At the same time as affording definite non-linguistic conceptions of conflict (as aspectual incompatibility) and progress (as relative accuracy, scope, and simplicity), the Gestalt Model provides a positive conception of incommensurability. The Gestalt Model thus differs from the Deductive Model in the above sorts of ways, and as will be seen below, it differs further in that it can function as the basis of a conception of theory relations which is capable of handling meaning variance between theories, and which provides a more realistic conception of the nature of scientific theories themselves.

CHAPTER 9

THE PERSPECTIVIST CONCEPTION OF SCIENCE

1. THE PERSPECTIVIST CONCEPTION AS BASED ON THE GESTALT MODEL

In being based on the Deductive Model, the Popperian and Empiricist views of science take scientific laws and theories to be statements of the form: for all x, if x has the property F, then x has the property G. The first step is conceptually to delineate a universe of x's having property F, and then it is to be empirically determined whether such x's also have property G. A number of these x's are thus to be observed and found either to have or not to have this property. There is no middle way—on this view either the predicate G is applicable or it is not, and if not, the law or theory is considered false.

In the present account quite a different tack is taken. First, a distinction is made between scientific laws and theories; and, with regard to *theories*, rather than start conceptually with a universe of objects which are to have some particular property, one may better be thought of as starting with some property, and then going on to find out in what sorts of instances it is applicable. In certain cases it may be discovered to fit quite well; in others, not so well. But the sharp distinction made in the logic of science between a predicate's either being applicable or not being applicable is here dropped.[1] On the present view a predicate may be more or less applicable, either by itself or as relative to some other predicate.

So here scientific theories are not taken to be statements having a truth value, but are instead likened to individual empirical concepts or predicates which are intended to apply to certain phenomena—predicates such as 'red' or 'round', or, with reference to the Gestalt Model, concepts such as 'picture-duck' and 'picture-rabbit'. Thus theories are here conceived of as being intended to apply to certain

[1] This difference thus serves also to distinguish the present view from the set-theoretic conception—cf. Chapter 11 below.

states of affairs, and to be such that they may be judged to be more or less successful in their application.[2]

The motivation for this alternative approach is the desire to provide a conception of science differing from the Empiricist and Popperian views in that it is able to account both for theory conflict and for scientific progress, and is also able not only to handle the problems of meaning variance and the nature of scientific theories, but to explain the way in which actual theories might well be incommensurable. Also, though the present view has gained much inspiration from the respective works of Kuhn and Feyerabend, it is not being presented as a direct reconstruction of the views of either of them; and, in fact, the notion of incommensurability to be treated here differs essentially from that described by Feyerabend as recently as in his *Science in a Free Society* (1978). Furthermore, the present conception is not directly concerned with the psychological or sociological aspects of science, but is to fall wholly within the realm of what is normally considered to be the philosophy of science. It might also be mentioned here that the present view is not at all intended to be prescriptive, or to provide 'methodological rules' for science or any other epistemological activity. Rather, it is intended to afford a realistic conception of the philosophically interesting aspects of science—a conception in which the move from one scientific theory to its successor is seen as being based on both rational and empirical considerations. Nor is the conclusion drawn from this that science itself provides the most reasonable means of gaining knowledge about reality, or that it is for any other reason a discipline which ought to be pursued. In this study such questions are left completely open.

The basis for the alternative conception to be presented below is, in this study, being taken as the Gestalt Model. It may be noted however that a different starting point might have been taken, such as, for example, a comparison of colour concepts, or a direct treatment of

---

[2] Cf. Kuhn (1962), p. 147: "All historically significant theories have agreed with the facts, but only more or less. There is no more precise answer to the question whether or how well an individual theory fits the facts. But questions much like that can be asked when theories are taken collectively or even in pairs. It makes a great deal of sense to ask which of two actual and competing theories fits the facts *better*."

scientific theories themselves.[3] It is because so many features of gestalt switch phenomena may be seen to have counterparts in the comparison of rival scientific theories that the heuristic value of these phenomena has resulted in their here attaining the status of constituting the basis for the present view. The following presentation will thus parallel the presentation of the Gestalt Model in the previous chapter, but certain points to be made will occasion the use of other sorts of examples and analogies.

2. SCIENTIFIC THEORIES AS CONCEPTUAL PERSPECTIVES

In the Gestalt Model the seeing of an aspect of a gestalt figure has been characterized as the application of a concept to the figure. In the present case then, a scientific theory will be understood as being in important respects analogous to a concept intended to have such an application. This notion of theory is to be as close to ordinary usage as possible, and thus should be in keeping with its definition in the *Concise Oxford Dictionary* as a "supposition or system of ideas explaining something, esp. one based on general principles independent of the facts, phenomena, etc. to be explained". From this definition we see that, in the terms being used here, not only should it be intended that a theory apply to certain states of affairs, but, in order to be a theory, it should actually be successfully applied. In this way then we might distinguish between mere hypotheses and actual theories. In what follows however, the term "theory" will be used in a broader sense so as to include what might otherwise be considered mere hypotheses, and emphasis will be placed on the idea of a scientific theory as an entity intended to have application to certain empirical states of affairs—an intention which may or may not be fulfilled.

---

[3] The present conception as a matter of fact originated independently of the Gestalt Model, and was introduced in an earlier study via an example in which the different systems of co-ordinates formulable within the special theory of relativity were each taken to constitute a 'conceptual perspective'. Thus not only might the particular account of science being presented here have taken a different starting point, but the Gestalt Model itself is by no means intended to function as the basis of a philosophy of a more general nature.

Of course scientific theories seem intuitively to be much more complex than individual concepts, and the notion of a single empirical concept is here meant only to be employed as an analogue to that of a scientific theory. But scientific theories and empirical concepts intended to apply to certain states of affairs have sufficient in common that they may be grouped and discussed under the one heading: *conceptual perspective*. Thus where accounts based on the Deductive Model take scientific theories to be statements of the form: $(x)(Fx \rightarrow Gx)$, here they are taken to be conceptual perspectives; and it is to the task of clarifying this notion, and of revealing in a general way its relevance to actual science, that the remainder of this chapter is devoted.

So an applied empirical concept may thus be thought to be a simpler form of conceptual perspective, and a scientific theory a more complex one. In the case of scientific theories a conceptual perspective may consist of a number of abstract notions, related in clearly defined ways. (In the next chapter the nature of the internal structure of scientific theories will be treated in greater detail.) More generally, a conceptual perspective of the more complex form is a sort of intellectual point of view, i.e. it is a conceptual system or framework which provides a structure in which one's thoughts about some particular aspect of the world can be organized. And while the interest of the present study is in scientific theories, there is no reason why we could not in a different setting consider, for example, such geometrical theories as those of Euclid, Lobachewsky, and Riemann each to constitute a conceptual perspective being applied to geometrical space.

With regard to the Gestalt Model we see then that the application of a scientific theory is to bear essential similarities to the phenomenon of seeing under an aspect,[4] and that just as the move from one gestalt concept to the other involves a fundamental and revolutionary change, so does the move from one scientific theory to its successor. Both may be said to involve a 'shift of conceptual perspective'—a

---

[4] Cf. Hanson (1958), p. 90: "Physical theories provide patterns within which data appear intelligible. They constitute a 'conceptual Gestalt'."

shift which does not preclude a meaningful comparison of the perspectives concerned.[5]

It may be thought though that where in the Gestalt Model one can easily shift back and forth between perspectives, this is not so in the case of scientific theories. But, while it may take some time before a scientist grasps or *understands* a new theory, the actual shift from not understanding to understanding, a step which may occur more than once before the new theory or perspective is internalized, is clearly similar to the dawning of an aspect which occurs in the Gestalt Model. It may be noted also that such an understanding is not gained as the direct result of argument or critical discussion,[6] but on the contrary requires an uncritical posture on the part of the individual seeking to understand. And, once the scientist has understood both theories, though he might believe one of them to be clearly superior to the other, he can nevertheless easily shift from working within one to working within the other, just as a person can easily shift from one aspect of the Gestalt Model to the other.

3. LOGICAL SIMULTANEITY

The Gestalt Model also provides a notion of simultaneity that is of relevance in the case of successive scientific theories. Thus, just as the concepts 'duck' and 'rabbit' cannot be applied simultaneously to the duck-rabbit, we should say that competing scientific theories cannot be applied simultaneously.

Other authors have employed a notion of simultaneity similar to that intended here. For example, in the context of discussing the nature of incommensurability, Feyerabend has done so in saying that: "we may suspect that the family of concepts centring upon 'material object' and the family of concepts centring upon 'pseudo-after-image'

---

[5] Cf. Kuhn (1976), pp. 190–191: "Most readers of my text have supposed that when I spoke of theories as incommensurable, I meant that they could not be compared. But 'incommensurability' is a term borrowed from mathematics, and it there has no such implication." For a recent instance of a misunderstanding of Kuhn in this regard, cf. Laudan (1977), p. 143.

[6] For a similar point of view, see Feyerabend (1975), p. 229.

are incommensurable in precisely the sense that is at issue here; these families cannot be used simultaneously . . . ."[7]

Part of what is of interest in considering the present notion of simultaneity is that it reveals a particular sense in which two distinct theories are to be independent of one another which is essentially different from the sense in which they might be conceived to be independent on a formal approach. From a formal point of view, a consideration of the relation between two theories, even if the relation is taken to be that of contradiction, requires that both theories be conceived as constituting the whole or part of one and the same formal system. But on the present conception, distinct theories are seen to be such that they cannot be applied simultaneously, and thus in principle to be such that, in application, they cannot be part of the same system. And this in turn means that when they are being treated as applied systems, formal relations cannot be established between them.

This being the case however does not preclude the determination of formal relations when the theories concerned are considered independently of their applications, i.e. when they are considered as purely formal systems. But it does suggest that the realization of such relations might well be misleading, if for no other reason than that it perforce ignores the fact that the systems are independent in the above sense. Thus, in keeping with the negative sense of 'incommensurability' given in Chapter 7, we should here say that, as *applied* systems, competing scientific theories cannot be treated *logically simultaneously*.

4. PERSPECTIVAL INCOMPATIBILITY

As is suggested by the Gestalt Model, on the present account theory conflict is not conceived to be essentially of a linguistic nature, nor to

[7] Feyerabend (1975), pp. 228–229. For another use of the term "simultaneity" in the context of the incommensurability question see Stegmüller (1979), p. 69.

The concept of simultaneity given above is also potentially applicable to the case of complementarity in quantum physics, as is suggested e.g. by P. Jordan's saying: "The properties connected with the wave nature of light on the one hand and those connected with its corpuscular nature on the other . . . can never appear in one and the same experiment at the same time" (1944), p. 132.

involve the express denial or negation of (part of) one theory by another. More generally, the sort of conflict of interest here—*perspectival incompatibility*—is independent of whatever formal, logical, or syntactical relations might hold between the theories; i.e. it is independent of how the theories are related to each other when considered simply as unapplied systems. Rather, theory conflict is here conceived to arise in the *application* of the theories, each of which is intended to provide a positive account of one and the same realm of phenomena, and this conflict is to involve each of the theories as a whole.

The Relevance of the Incompatibility of Colour Concepts

As outlined above, both single empirical concepts and more complex empirical systems, when applied, constitute conceptual perspectives. In the case of empirical concepts there exists a well known situation in which conflict arises on the basis of two positive accounts being given of one state of affairs. And such conflict, in the final analysis, is not a logical or formal one. It is the conflict that arises in the simultaneous application of different colour concepts. In this regard Bertrand Russell has said:" "this is red" and "this is blue" are incompatible. The incompatibility is not logical. Red and blue are no more *logically* incompatible than red and round. Nor is the incompatibility a generalization from experience."[8]

It has been suggested by D. J. O'Connor that the lack of formal conflict here is a result of the fact that " 'red' and 'blue' and terms like them ... can be defined only ostensively".[9] And it has been thought by a number of philosophers that such incompatibility is restricted only to a small range of empirical concepts which are dependent on ostension for their meaning, including perhaps certain shape concepts, such as 'round' and 'square'. Other empirical concepts, they believe, can be analyzed so as to reveal an underlying contradiction, which may be taken to account for their conflicting.

But if we take as the criterion for a concept's being empirical that its definition ultimately rest on ostension, then, as has been argued by

---

[8] Russell (1940), p. 100.
[9] O'Connor (1955), p. 112.

Lionel Kenner, "the difficulty which is seen with colour words applies, in the end, to all empirical concepts."[10] Thus, though the considerations of the above authors have not led them to the distinction made in the present study between the application of an empirical concept and the claim that the concept applies, Kenner's argument nevertheless suggests that even in the latter case the nature of the incompatibility between such concepts would not be logical or formal. And on the present view, where we are not concerned with the making of statements but only with the application of concepts, we should say that in their application to the whole of one and the same thing, distinct empirical concepts might well be perspectivally incompatible.

Formal or Syntactical Contradiction vs. Contradiction Proper vs. 'Contrariety'

In order to clarify the view taken here it may prove helpful to distinguish between certain sorts of incompatibility which are candidates for characterizing the nature of the conflict occurring in the case of successive scientific theories. To this end then we might begin with the notion of formal, syntactical, or logical contradiction. Such a 'contradiction' manifests itself in the syntax or form of certain linguistic expressions—more particularly, in the form of *sentences*; and it involves the conjunction of a sentence with its *negation*. Thus we should say that the (molecular) sentence "This is red (all over), and it is not the case that this is red (all over)" is a formal contradiction. The essence of this formal contradiction can be captured by formulating the above sentence in the sentential calculus as: $P \& \neg P$, or in the predicate calculus as: $Pa \& \neg Pa$. Thus the nature of the conflict between competing scientific theories as conceived on the Deductive Model is of this sort.

Formal or syntactical contradiction may be distinguished however from contradiction proper in that where the former is a relation between sentences and involves negation, we may say that the latter is a relation between *statements* or, perhaps better, between *proposi-*

---

[10] Kenner (1965), p. 151. Kenner however takes this to imply the *triviality* of the problem with colour concepts, while it is here seen as suggesting its *generality*.

tions, and involves *denial*. Thus, to take an extreme example, if a person were to say "This is red (all over) and it is not the case that this is red (all over)", and meant a different thing by "this" or "red" on each of their uses, while we would have a syntactical contradiction, we would not have a contradiction proper. (This is essentially the problem of meaning variance.)[11] While sentences may be used to express propositions, the form of the sentence is not itself a guarantee that the proposition it is being used to express is or is not a proper contradiction.

A third notion of relevance to the present case will here be called 'contrariety', and may be exemplified by a statement such as 'This is both red (all over) and blue (all over)'. Contrariety thus differs from contradiction proper in that its faithful linguistic representation does not, prima facie, evince a syntactical contradiction, and in that it necessarily involves the employment of 'positive' alternatives, and thus cannot be simply a case of affirmation and denial. Thus, for example, the formulation of the above statement in the predicate calculus would give: $Pa \;\&\; Qa$; and this syntactical representation fails to capture the conflict involved in the statement it represents. Furthermore, as is evidenced by this formulation, where it should always be the case that one of two contradicting statements be true, both of two contrary statements may be false.

The notion of perspectival incompatibility thus comes closest to what has here been called 'contrariety', but differs from it in the essential respect that it does not involve statements or propositions having a truth value, but involves rather the application of concepts. Thus where the propositions 'This is red (all over)' and 'This is blue (all over)' evince 'contrariety', the attempt simultaneously to apply the concepts 'red' and 'blue' to the whole of one thing constitutes an instance of perspectival incompatibility.

Conflicting Perspectives Suggesting the Same Results

It may be emphasized that perspectival incompatibility arises in such cases independently of there being any sort of anomaly favour-

[11] For a (very) early discussion of this problem which is quite in keeping with the present view see Campbell (1920), pp. 51 ff.

ing one perspective at the expense of the other, and that two concepts being perspectivally incompatible does not in itself imply that either of the concepts taken alone is not applicable to the state of affairs in question. At the level of simple empirical concepts this may be made clear by considering Wittgenstein's cube gestalt switch.[12] The drawing cannot be seen as two sorts of cube at the same time—in this way the two cube concepts are perspectivally incompatible; nevertheless, unlike the case involving the duck-rabbit, the drawing is not itself more in keeping with one of the concepts than with the other.

In considering actual science a situation similar to this arises, for example, in the case of the Copernican and Tychonic (note: not Ptolemaic) systems of astronomy.[13] Each system conflicts with the other in constituting its own conceptual perspective on the motion of the planets, but both systems suggest their having the same motions relative to one another and to the sun. Thus, in application to the solar system, we might say that these systems are not incompatible at the level of observation, while they are incompatible at the level of theory. Another example of this sort is provided by the caloric and dynamical theories of heat, as described by Kuhn: "In their abstract structures and in the conceptual entities they presuppose, these two theories are quite different and, in fact, incompatible. But, during the years when the two vied for the allegiance of the scientific community, the theoretical predictions that could be derived from them were very nearly the same".[14]

The existence of these sorts of cases in science leads directly to a questioning of the Popperian conception of theory conflict, for on Popper's view theory clash is to *depend* on the theories concerned giving conflicting results, such that if both of two theories suggest the same results they ought not be considered as being incompatible. Popper says: "in order that a new theory should constitute a discovery or a step forward it should conflict with its predecessor; that is to say, it should lead to at least some conflicting results. But this means, from a logical point of view, that it should contradict its predeces-

---

[12] Wittgenstein (1921), § 5.5423.

[13] Cf. Kuhn (1957), pp. 201 ff. For a discussion relevant to the present one, see Hanson (1966).

[14] Kuhn (1961), pp. 176–177.

sor".[15] This picture is of course quite in keeping with Popper's view as based on the Deductive Model, but it means that he is unable to account for the conflict that may arise in those cases where the theories concerned do not differ with regard to their predictions. But this in turn suggests that the situation is not simply that Popper's conception of theory conflict is inapplicable in a few odd cases, but rather that even in those cases where distinct theories do suggest incompatible results, this difference between them is not essential to their conflicting; and if this is so, then Popper's notion of theory clash does not capture the essence of the nature of theory conflict in *any* case in which it occurs.

Concerning the Problem of Meaning Variance

The difficulty of applying the Popperian concept of theory conflict to the case of actual theories has been treated earlier with regard to the problem of meaning variance. Here we will look at this problem somewhat more closely, and show how it can be overcome on the present conception. An early formulation of the problem is given by Dudley Shapere, though he sees it as being a problem for Feyerabend's view rather than for those philosophies of science based on the Deductive Model. With regard to Feyerabend, Shapere asks:

> How is it possible to reject *both* the consistency condition *and* the condition of meaning invariance? For in order for two sentences to contradict one another (to be inconsistent with one another), one must be the denial of the other; and this is to say that what is denied by the one must be what the other asserts; and this in turn is to say that the theories must have some common meaning.[16]

If we approach this problem first on the level of basic epistemology, and take the meaning of a (descriptive) term to be the (empirical) concept it is being used to express, earlier considerations suggest that in certain cases conflict can occur precisely *because* there is a change

---

[15] Popper (1975), pp. 82–83.

[16] Shapere (1966), p. 57. See also e.g. Achinstein (1964), p. 499, and Giedymin (1970), p. 265. Note also Stegmüller's remark that, as criticisms of the views of Kuhn and Feyerabend, "*Almost all* of Shapere's arguments are based on the statement view and, for the most part lose their force with its rejection." (1973), p. 261.

in the meaning of such a term.[17] If one person says that something is red, and is using the term "red" as it is normally used, while another person says that the same thing is red, but in saying so actually means 'blue' by the term "red", then we have an instance of the sort of conflict earlier called 'contrariety'.

Or, to take an example from the history of language, we might consider the English term "nice". A few centuries ago this term had a meaning 'wanton' or 'lascivious', and it has now come to mean 'kind, considerate, pleasant to others'.[18] If we grant that a person cannot be lascivious and at the same time kind and considerate, we can see how the change in the meaning of "nice" would *account* for the conflict involved in saying of a person that he is nice, meaning 'lascivious', and that he is nice, meaning 'kind'.

This same reasoning can be applied to the much discussed issue regarding the meaning of the term "mass" in the context of Newton's and Einstein's gravitational theories respectively. As used in the context of Newton's theory, implicit in its meaning is the idea that the mass of a body is independent of its velocity. The same term, employed in the context of Einstein's theory, implies on the contrary that the mass of a body increases with its velocity. In this way then we might say that a change in the meaning of the term "mass" would account for the conflict involved in saying of one and the same moving body that it has some one particular mass both in Newton's sense and in Einstein's.

The above examples are intended mainly to suggest that different statements may be inconsistent even when the descriptive terms common to them have different meanings (senses). As has been mentioned above however, on the present conception theory conflict is not viewed as being based on the inter-theoretical relations among the meanings of individual terms, but is seen rather as resulting from the attempt to apply the whole of certain distinct theories to the same state of affairs. Thus while the present view allows for the meaning variance of individual terms, and while the differences in the particular natures of the respective theories might in certain cases be seen as

[17] M. Hesse touches on this point and provides an example relevant to it in her (1963), pp. 102–103.
[18] Ullman (1962), p. 234.

resulting from the sum of these sorts of conceptual differences, it is the theories as wholes that are here seen to conflict. In other words, on the present view incommensurability between theories involves a shift of conceptual perspective—a shift which may or may not be accompanied by a meaning change of individual terms.

The preceding discussion thus provides a means by which Shapere's criticism of Feyerabend with regard to meaning variance and incompatibility can be overcome. In spite of this however, unless Feyerabend is willing to allow that two incommensurable theories can both pertain to one and the same state of affairs, his view of incommensurability is still open to charges of relativism which the present conception succeeds in avoiding.

5. INCOMMENSURABLE THEORIES HAVING THE SAME INTENDED DOMAIN

Throughout his writings in the philosophy of science Feyerabend has maintained a form of realism—'epistemological realism'—which suggests that in accepting a scientific theory one is to take the theoretical constructs appearing in the theory as having real counterparts with the properties predicated of them by the constructs. Thus, on Feyerabend's view, if two theories have in their respective bases essentially different constructs, then the entities in the world to which these theories apply, if they apply at all, are themselves to have essentially different properties. And, so the reasoning goes, seeing as such entities are referred to, if not solely then primarily, by those constructs in terms of which they are characterized in such theories, then theories based on essentially different constructs ought to be taken as referring to essentially different things. We might in fact say that on most occasions where Feyerabend speaks of theories being incommensurable, he has in mind that they differ in this sort of way. Thus we read, for example, in his *Science in a Free Society*:

> ... All we need to do is to point out how often the world changed because of a change in basic theory. If the theories are commensurable, then no problem arises—we simply have an addition to knowledge. It is different with incommensurable theories. For we certainly cannot assume that two incommensurable theories deal with one and the same objective state of affairs (to make

the assumption we would have to assume that both at least *refer* to the same objective situation. But how can we assert that 'they both' refer to the same situation when 'they both' never make sense together? Besides, statements about what does and what does not refer can be checked only if the things referred to are described properly, but then our problem arises again with renewed force.) Hence, unless we want to assume that they deal with nothing at all we must admit that they deal with different worlds and that the change (from one world to another) has been brought about by a switch from one theory to another.[19]

With regard to this line of reasoning we note that even on its weaker reading (in which the world does not change) it creates difficulties when it comes to explaining the experiments in actual science (e.g. Michelson-Morley) which have been considered crucial with respect to the viability of the sorts of theories Feyerabend takes to be incommensurable. If incommensurable theories concern different worlds—are about different things—then how are we to conceive of such theories as competing, or winning or losing in their competition? Feyerabend's 'pragmatic theory of observation', for example, does not answer this question, for it suggests that each theory has its own criteria of acceptability.[20]

On the present view however this relativistic consequence is avoided, for here we should say that when a scientist moves from one of two incommensurable theories to the other, he can still very well *intend* that both theories apply to the same states of affairs, even if they characterize those states of affairs in essentially different ways. Furthermore, on the basis of certain criteria (such as the performance of the same sorts of operations), we, as onlookers, can often judge that the scientist is treating of one and the same aspect of reality in his respective applications of the two incommensurable theories.

In fact, in *Against Method*, Feyerabend has himself discussed the notion of incommensurability in such a way as can be naturally extended to this view of the situation. In keeping with the present conception as based on the Gestalt Model, as mentioned in Chapter 7 above he has suggested that gestalt switch figures provide instances

[19] Feyerabend (1978), p. 70.

[20] For a discussion of Feyerabend's 'pragmatic theory', with relevant references, see Shapere (1966), pp. 59ff. See also Feyerabend's discussion of crucial experiments in e.g. his (1970), pp. 226ff.

of incommensurability—and in such cases it is clear that the incommensurable concepts concerned are both being referred to the same thing, namely, the gestalt figure itself. And, in another place, Feyerabend expresses a view similar to the present one in saying that:

> When we go from classical physics to relativity, what remains the same are the objects. The objects are what they are, only we think different things about them. To a large extent, our operations for getting certain numbers remain the same. The functors which we use and much of the syntactical apparatus may also remain the same. What is different are all the concepts connected with the functors.[21]

Thus, in keeping with this line of thought, it is here suggested that even on a realist conception incommensurable theories might well pertain to the same states of affairs, or have the same *intended domain*. Of course there exist theories which are intended to treat of different things, and, if one wanted, one might say of such theories that they are incommensurable. But they would not be incommensurable in the interesting sense treated here, where the theories concerned conflict with one another.

Speaking more generally, we should here say that differing conceptual perspectives can have the same *reference*.[22] The use of this term is to suggest the idea that a conceptual perspective, in being an applied concept or system of concepts, in a sense 'points' in a certain direction. And the direction in which it 'points' is in turn dependent upon the *intention* of the person applying the concept or conceptual framework, and not on the concept itself. Thus, as has been suggested in the previous chapter, the 'duck' and 'rabbit' concepts of the Gestalt Model might each be intended to apply to a different figure, and consequently be said to be given different references. It may also be noted that where a concept or system of concepts is *given* a certain reference, a conceptual perspective, in consisting of an *applied* concept or framework, *has* a particular reference. And it is only when two perspectives have the same reference that they conflict.

Furthermore, it may be pointed out that in order for such conflict to arise it is not the case, on the present view, that that toward which the

[21] Hanson et al. (1970), p. 247.

[22] In Dilworth (1978) the term "intention" was used to cover both of the notions 'intention' and 'reference' treated in the present study..

perspectives are directed need actually exist. (The duck-rabbit figure might be a mirage—but even if it were, the 'duck' and 'rabbit' concepts could not be applied to it simultaneously.) The conflict is in this way dependent on the intention(s) of the person(s) employing the concepts, and not on the existence of that to which it may be believed that they are being applied. Thus in the case of scientific theories for example, we might have two different theories which conflict with one another as regards the nature of certain sub-atomic particles —particles which further research reveals not to exist at all.

6. SYSTEMS INVOLVING THE SAME CATEGORIES

In the present conception the notion of commonality of reference affords a way of conceiving how two radically different theories might well be about the same realm of phenomena, or have the same intended domain. In order for them to conflict though, they must also involve the same *categories*.

The individual predicates 'red' and 'blue', as versus the predicates 'red' and 'round', are of the same category, 'colour', just as in the Gestalt Model the concepts 'picture-duck' and 'picture-rabbit' may be thought to be of the same category 'shape'. Scientific theories however, in that each constitutes a more complex sort of conceptual perspective, rather than simply being of just one category, may each be thought to involve a number of categories. And, in considering more than one theory, the categories they involve may or may not be the same.

The distinction between the concept of category and that of class or set helps mark a difference between the present account and those of the Empiricists and Popperians. Allowing that predicates can be grouped according to their category, and that things may be grouped into classes depending on the applicability of certain predicates, on the Empiricist and Popperian views scientific laws and theories may most simply be seen as claiming that one class or set of things is a subset of another, that is, that the class of all things x to which a particular predicate F is applicable is a subset of the class of all things y to which a particular predicate G is applicable. And conflict is thought to arise when a further claim is made to the effect that there exist things in class F which are not in class G.[23]

Now this approach gives rise to basic ontological, epistemological, and logical problems, a detailed discussion of which lies beyond the scope of the present study. These problems include, for example, how we are to determine what is to constitute a thing in our universe, and what we are to do in cases where it is debatable whether some such thing really does or does not possess some particular property, as well as what steps are to be taken so as to avoid the possibility of formulating the classical logical paradoxes which arise from the unrestricted use of the principle of abstraction, whereby a class or set is defined on the basis of its elements having a certain property.

On the present account however, in keeping with the view of Evandro Agazzi, the objects of which science treats are given by certain *operations*; or, in the case of physics, they consist more particularly in the *data* resulting from certain *measurements*. Thus we follow Agazzi where, for example, he suggests that we

> [consider] what the different sciences do in order to treat 'things' from their 'viewpoint': they submit them to certain specific manipulations of an *operational* character, which put the scientist in the position of answering certain specific questions he can formulate about these things. Such operational procedures may be the use of a ruler, of a balance, of a dynamometer, in order to establish some physical characteristics of the 'thing' like its length, its weight or the strength of some force exerted on it; they may be the employment of some reagents to determine its chemical composition, etc.[24]

In the same vein then, we should here say that evidence suggesting the particular categories with which a certain theory is involved may be obtained via a consideration of the sorts of operations or measurements used in applying or testing the theory. Thus two distinct theories' pertaining to the same measurements suggests that they involve the same categories.[25]

[23] For a similar description of this approach, and a discussion of it, see Wartofsky (1968), pp. 277 ff.

[24] Agazzi (1976), p. 148. Note that Agazzi's 'viewpoints' are here more closely aligned with different sciences than with different theories. For a development of this view see the rest of Agazzi (1976), as well as Agazzi (1977 *a*).

[25] Kuhn's view of incommensurable theories thus differs from the present one to the extent that he suggests that such theories need pertain to different data; in this regard cf. Kuhn (1962), p. 126, and (1974), p. 473 n.

The above may be brought out by considering the application of colour concepts. It may be wondered, for example, whether some particular thing is more red than orange, or vice versa, i.e., whether one of the predicates 'red' or 'orange' is the better applicable. The scientific approach would be to take a reading on the thing in question using a spectrometer.[26] And the fact that the same operation is of relevance to the applicability of both predicates suggests that both are of the same category (viz., 'colour').

The distinction between theories which are perspectivally incompatible and those which are not may in turn be made clearer through a consideration of the following simple schema.

Category A 'colour'	Category B 'shape'	Category C ...
1. 'red'	1. 'round'	1. ...
2. 'blue'	2. 'square'	
3. ...	3. ...	

Table 1.

We may say that the determination of the applicability of the predicates in the category 'colour' involves the use of a spectrometer, and the applicability of those in the category 'shape' is determined by the use of a ruler and compasses. That different operations should be performed implies that the predicates are in fact of different categor-

[26] Cf. a slightly different example given in Allen & Maxwell (1939), p. 5: "We can better understand the difficulties of the earlier investigators if we consider the question of *measuring* colour.

Colour we generally regard as a quality. It is, however, possible to select a scale in which a particular colour, blue for example, is graduated in depth from very pale to very dark. We could go further and, by making a *mental estimate*, attach a series of numbers to the various samples so that we could speak of any given sample of blue as having so many "degrees of blue". This is not quite the same process as a *physical measurement*, but we might take a further step and try to find a correlation between our mental scale and a physical scale derived, for example, from the amount of dye used in preparing a specimen or from the relative proportions of the blue and white sectors in a rotating disc."

With regard to the relation between operations and scientific objectivity in the context of colour concepts see Agazzi (1978), pp. 100ff.

ies. And the attempt to apply, for example, the predicates 'red' and 'round' to the whole of one thing would not result in conflict. In being of different categories, 'red' and 'round', as applied predicates or concepts, would not be perspectivally incompatible.

A situation like this can occur in the case of scientific theories. One theory might be concerned with the motion of certain bodies, while another might consider those bodies solely from the point of view of the electromagnetic radiation being emitted from them. To the extent that the operations performed in the application of the two theories should differ, the theories would not be perspectivally incompatible.

Agazzi has made a point similar to the above, and has extended this way of thinking to provide a criterion for demarcating different sciences:

> [A]ny science delineates and determines its object sphere by means of a finite series of basic predicates. Such predicates are always defined in terms of operative criteria and they serve to provide this science with its objects. For instance, something is to be regarded as an object of mechanics if when speaking about it we use as basic concepts only concepts of mechanics (such as mass, space, time, force). ... Thus, we could say that any science determines through its basic predicates its own *whole* scope so that everything outside this scope is of no interest to it and lies outside its frame of reference. For example, electromagnetic phenomena are not part of mechanics because they cannot be dealt with in terms of the basic mechanical concepts.[27]

While two distinct theories would not be perspectivally incompatible if they involved the application of predicates from different categories, they could be incompatible if they respectively involved the application of predicates from within the *same* category (to the same intended domain). Where, for example, the predicates 'red' and 'round' can both be applied to the same thing, the predicates 'red' and 'blue' (or 'round' and 'square') cannot. The case is similar with different scientific theories: if they are related to the same operations, and consequently concern the same categories, then, other things being equal, they would be perspectivally incompatible, independent-

[27] Agazzi (1977*b*), p. 166. This quotation also has direct relevance to the tables appearing at the beginning of the next chapter.

ly of whether they suggested the same or different results of those operations.

7. RELATIVE ACCEPTABILITY: ACCURACY, SCOPE, AND SIMPLICITY

The presentation of the Gestalt Model in the previous chapter reveals three sorts of factors that may be taken into account in considering the relative acceptability of conflicting conceptual perspectives: accuracy, scope, and simplicity. Here, competing scientific theories, in constituting conceptual perspectives, are also to be thought of as comparable with regard to these factors, and on the basis of such a comparison may be judged as to their relative acceptability; and we can say that *progress* has been made when a theory is adopted which is, in terms of these factors, relatively more acceptable than its predecessor.

Accuracy

The example involving colour concepts used above to indicate the relevance of categories to the present view also shows the way in which one of two competing theories may be judged on the basis of measurements to be more accurate than the other. Thus, the readings obtained on a spectrometer might show one of the predicates 'red' or 'orange' to be the better applicable to some particular thing. And the same may be the case with regard to scientific theories: measurements which are relevant to the applicability of two competing theories may show one of them to be the more accurate—even in the case where neither theory suggests exactly those results obtained. Thus on the present view, while such theories would be *incommensurable* in that the move from one to the other involves a shift of conceptual perspective, they might nevertheless be considered *commensurable* in the sense that, and to the extent that, the same operations are employed to determine their respective applicability. And, where competing theories need not suggest different results in order to conflict (consider the example of the respective astronomical systems to Tycho and Copernicus), they must do so in order for one of them to be judged as being more accurate than the other.

But the theories' suggesting different results is not in itself sufficient to compare them in this regard. The results they respectively suggest must also differ sufficiently so as to lie outside of the range of mensural error of the instruments employed in the comparison. Thus the testing of certain theories as to their relative accuracy might have to await the development of sufficiently sensitive measuring instruments; and the failure to develop such instruments would preclude this sort of comparison.

Also, competing theories may for the most part suggest respective results which do not lie outside of this range of mensural error, but which are measurably different only when the theories are applied to a particular special case. In such a case then we can employ measuring instruments so as to stage a *crucial experiment*, the results of which may be taken to suggest one of the theories as being the more accurate.

Scope

An independent factor which may play a role in a debate concerning the relative acceptability of incommensurable theories is the *scope* or *generality* of each of them. Thus, in the simplest case, where two such theories are equally accurate in application to the intended domain they have in common, and one of the theories is also more or less applicable to states of affairs to which the other does not apply at all, then, other things being equal, it would have the greater scope of the two and would for this reason be the more acceptable. Of course there is nothing to prevent more complicated situations from arising in which, for example, one theory has a wider scope than the other, but the other is the more accurate in application to the domain to which both more or less apply. And it is clear that in such a case the debate over which of the two theories is all in all the more acceptable cannot be so easily resolved; and both theories might be retained, each to be used in those particular cases where the sort of superiority it evinces can come to the fore.

Simplicity

Where the notions of accuracy and scope indicate empirical factors which may play a role in the possible argumentation concerning the relative acceptability of competing theories, the notion of *simplicity* indicates a potential factor which is of a more rational nature. As is suggested by the Gestalt Model, the relative simplicity of a scientific theory may be likened to the ease or naturalness with which a gestalt diagram can be seen in one aspect as compared with another. Following the model further, we may say that the simpler of two theories is the one involving the fewer *ad hoc* modifications in its attempt to explain the relevant empirical data, i.e., the one which is the less *sophisticated.*

An example of relevance to this notion of simplicity is provided by a consideration of the astronomical systems of Ptolemy and Copernicus. This example is particularly suitable here, for the two systems have the same scope, and may be taken to be equally accurate. But in terms of its basic concepts Copernicus' system can give a good qualitative explanation of such states of affairs as the retrograde motion of the planets, the velocities of the planets in their orbits, the fact that Mercury and Venus never stray far from the sun, and so on. The basic functioning of these sorts of phenomena is *integral* to Copernicus' system, and the irregularities they evince are simplified and shown only to be apparent. On Ptolemy's system however, in its simplest form, these phenomena find no place at all, and it is only through ad hoc modifications that they can be accommodated. And where on Copernicus' system their apparent irregularity is shown to be illusory, on Ptolemy's it is represented by an irregularity in the system itself. Thus the planets do not only appear to loop back in their orbits, they are conceived as actually doing so; each planet does not only appear to move with an inconstant angular velocity as relative to the centre of its orbit, but is thought actually to behave in this way; and the handling of the fact that Mercury and Venus are never seen far from the sun is, on the Ptolemaic system, entirely ad hoc, and consequently does not explain this phenomenon at all.[28]

Now this is not to say that the full-blown version of Copernicus' system is free from sophistications, for as we realize today the orbits of the planets are more nearly elliptical than circular, with the sun standing at one of the foci of the ellipse described by each planet (on

---

[28] As regards this last point see Kuhn (1957), pp. 172–173. For the other points see the same work, esp. Chs. 2 & 5.

top of this it seems that a number of the modifications Copernicus made to his original scheme were occasioned by his working from erroneous data). Thus even Copernicus' system, in order to achieve an increased degree of accuracy, required ad hoc modifications; and the resulting scheme might well look as complicated as Ptolemy's. But the above considerations nevertheless suggest Copernicus' system to be the more coherent, and to be essentially simpler than Ptolemy's in the sense intended here.[29] We might say that it provides a conception of the motions of the planets which is easier to grasp 'in a single gestalt'.

Thus we see that on the present view not only is theory conflict explainable as perspectival incompatibility, but the problem of meaning variance is overcome, and a positive conception is given of the sorts of factors involved in determining whether one scientific theory should constitute a progression beyond another. In the next chapter the present view will be further developed in the context of an example taken from science, and it will there be seen to handle a number of other notions of philosophical interest as well, including those of idealization and theoretical terms, and it will also be seen to provide a more realistic conception of the nature of scientific theories than that afforded on the basis of the Deductive Model.

---

[29] Thus the notion of simplicity discussed here is a relative one, and the present explanation should answer Lakatos' question as regards how Copernicus' theory is simpler than Ptolemy's: see Lakatos (1970), p. 117 &n. Feyerabend too raises questions in this regard, but seems to admit the relative simplicity of the Copernican system when he asks: "Why should astronomers in the 16th century have accepted a physically and theologically impossible theory just because of its simplicity?" See Feyerabend (1978), p. 47.

CHAPTER 10

DEVELOPMENT OF THE PERSPECTIVIST CONCEPTION IN THE CONTEXT OF THE KINETIC THEORY OF GASES

1. PARAMETERS AS QUANTIFIED CATEGORIES

The conception of science and scientific progress presented in the previous chapter may be further explicated with the help of an example taken from the physics of gases. Though the presentation of this example will for the most part follow the actual development of gas theory, it is not intended to constitute the basis of an historical analysis, but to be a coherent reconstruction capturing the essence of the conceptual moves in this development. As a first step in the presentation of the example, the sort of schematization provided by Table 1 (p. 94) is here given a more definite form as a table of particular *parameters*, or quantified categories; and the general remarks made in the context of Table 1 should also be applicable here.

	Parameter A mass	Parameter B length	Parameter C time	Parameter D temperature
Value scale	$\mathbb{Q}$	$\mathbb{Q}$	$\mathbb{Q}$	$\mathbb{Q}$
Unit	kilogram	metre	second	degree Kelvin
Measuring instrument	balance	metre stick	clock	thermometer

Table 2.

In comparison with Table 1, predicates or concepts falling under a certain category are here rational number *values* of a parameter, expressed in terms of the *unit* specific to the parameter. And with each parameter is associated a measuring instrument, the appropriate use of which should allow the determination of the value of the parameter in a given empirical situation.[1]

From the parameters appearing in Table 2 we can obtain further parameters which are of direct relevance to the kinetic theory of gases:

	Parameter A′ volume	Parameter B′ pressure	Parameter D temperature
Value scale	$\mathbb{Q}$	$\mathbb{Q}$	$\mathbb{Q}$
Unit	cubic metre	newton per square metre	degree Kelvin
Measuring instrument	metre stick	manometer	thermometer

Table 3.

The parameters volume and pressure appearing in Table 3 are each derived from parameters appearing in Table 2. Volume is derived from length, and pressure is derived from mass, length, and time via the intermediate parameter *force*; and both volume and pressure are assumed to take rational number values.[2] The unit *newton* is that of the parameter force, and is defined as kilogram times metre per second squared. The parameter temperature is carried over directly from Table 2.

2. BOYLE'S LAW

In terms of Table 3, *Boyle's law* (1661) states that the (value of the) pressure times the (value of the) volume of a given gas, as measured by a manometer and metre stick respectively, is constant, given

[1] The standardization required in order for this and the following table to be applicable to a development spanning more than two hundred years has been facilitated by the employment of notions of contemporary science: e.g. those of *newton* and *degree Kelvin* appearing in Table 3 below. Also, following standard notation, quotation marks are not being used in referring to individual parameters. Nevertheless, parameters, as quantified categories, are not here conceived as existing in the world, but rather as being abstractions we employ in our attempts to understand it. Cf. footnote 18 to Chapter 8.

[2] For a description in which parameters can take real number values see the text to footnote 10 below.

constant temperature, as measured by a thermometer.[3] This may be abbreviated to:

(28) $pV = R$ (given constant temperature).

As presented here, Boyle's law constitutes what physicists would call an *empirical* or *experimental* law. While it does express a regularity of nature, the law itself is based solely on results obtained using instruments in actual measurements, and is not here related to theory. It does not succeed in explaining why, given constant temperature, the volume of a given mass of gas should be inversely proportional to its pressure; it tells us only that this is generally the case.

In application to real gases Boyle's law does not hold exactly, giving particularly divergent results from those obtained by measurement in the case of gases at high pressure and low temperature, i.e., when the gas in question approaches that critical point at which it begins to condense into a liquid. A gas to which Boyle's law applies exactly may thus be defined as an *ideal gas*. A formula subsuming Boyle's law, involving the further realization that the product pV in Boyle's law varies proportionally to the temperature (T), constitutes the equation of state for an ideal gas:

(29) $$\frac{pV}{T} = R.$$

This same equation, when applied to *real* systems, is called the *general gas law*, and serves to express the relation expected to obtain among the values of all three of the parameters volume, pressure and temperature.

3. THE IDEAL GAS MODEL

A model for an ideal gas has been developed in the context of the kinetic theory of matter (Bernoulli, 1738).[4] According to the model as

[3] In its original formulation, Boyle's law did not involve the parameter temperature. In subsequent developments however it was realized that the applicability of the law requires that temperature be held constant.

[4] Bernoulli is generally recognized as being the first to suggest a model of the sort which is today called the ideal gas model. It may be noted however that in Bernoulli's model there is to be an infinite number of molecules: cf. Partington (1961), p. 477.

understood today, an ideal gas consists of molecules in motion, and has the following further properties:

1. The volume of all the molecules taken together is negligible in comparison with the volume of the container occupied by the gas;
2. there are no forces acting on or between the molecules except in the case of collision;
3. when colliding with each other or with the walls of the container the molecules act as completely elastic spheres;
4. the time a molecule takes in colliding is negligible as compared with the time between its collisions;
5. the motion of the molecules is completely random.

Boyle's law can be derived from the model for an ideal gas. The derivation involves assuming, among other things, that a particular sample of the gas contains some definite number of molecules each with the same mass and all of which share a certain average velocity.[5] We can also derive from the model the experimental law of Gay-Lussac (1808) concerning the relation between the volumes of interacting gases in the case where chemical changes occur. When so derived, such experimental laws are said to be *explained* by the model (or by the theory in which the model plays a central role). We can also say that the model explains, for example, the pressure of a gas, as manifest in manometer readings, as being the result of molecules bombarding the walls of the container.

As mentioned above, Boyle's law, or the general gas law, fails to give values close to those obtained via measurement at high pressures and low temperatures. More particularly, it gives no hint that, when a gas at a temperature below a certain critical point is highly compressed, it will undergo a significant decrease in volume and become a liquid.

One way to account for this decrease in volume on the kinetic theory of gases is to assume that there is an attractive force operating between the molecules which becomes effective when the molecules are in relatively close proximity to one another (i.e. when the gas is under high pressure), thus sharply repressing their motion at a par-

---

[5] This is the approach taken e.g. in Barton (1933), pp. 197–201. It seems however that the derivation should also be possible even assuming an infinite number of molecules, as in the case of Bernoulli's model: cf. Partington (1961), p. 477.

ticular critical point. Also, it may be noted that while it is understandable that at low pressures the total volume of the molecules of a gas may be considered negligible as compared with the volume of the vessel containing the gas, as the gas is compressed this assumption becomes moot.

But both of these notions—that of an attractive force operating between the molecules and that of their having a non-negligible volume—run counter to the ideal gas model. Thus if these factors are to be taken into account in attempting to explain the behaviour of gases under high pressures, as well as the phenomenon of change of state, a new model must be devised.

4. VAN DER WAALS' LAW

Still in accordance with the kinetic theory of matter, *van der Waals* (1873) developed a model capable of being used to explain the behaviour of gases under high pressure, as well as their changes to the liquid state. In the model inter-molecular forces are present, and molecules are assigned a definite volume. The equation of state of van der Waals' model may be presented in the following form:

$$\left(p+\frac{a}{V^2}\right)(V-b)=RT, \tag{30}$$

where a and b are constants characteristic of the gas in question. More particularly, the expression "a/V^2" represents the decrease in pressure (as expected on the equation of state of the ideal gas model) exerted by a substance as a result of the effect of inter-molecular forces; and the term "b" represents the effected decrease in the molecule-free volume of the container as a result of the volume of the molecules. For $a=b=0$, (30) reduces to the equation of state for an ideal gas.

As an expression of van der Waals' *law*, (30) is thus to have application not only to substances in their purely gaseous state, but also to substances as they undergo a transition from their gaseous to their liquid form.

If the pressure, volume, and temperature are expressed as fractions of the *critical* (i.e. 'change of state') pressure, volume, and temperature, which are constants, we obtain van der Waals' *reduced equation of state*:[6]

$$(31) \qquad \left(\mathcal{P}+\frac{3}{\mathcal{V}^2}\right)\left(\mathcal{V}-\frac{1}{3}\right)=\frac{8}{3}\,\mathcal{T}.$$

This equation does not contain the constants a and b, which differ for different gases, and so is a generalized expression which applies to all gases to which the individual equations having the form of (30) apply.

While van der Waals' equation does not succeed in giving exactly correct results, it nevertheless constitutes an advance over the general gas law in that it finds application to the phenomenon of change of state, in most cases providing results of at least the correct order of magnitude.

5. EXPLICATION OF THE PERSPECTIVIST CONCEPTION IN THE CONTEXT OF THE EXAMPLE

In the previous chapter the basic notions of the present conception were presented and discussed against the background of the Gestalt Model, and involved the employment mainly of simple empirical predicates such as 'red' and 'round'. The central notions treated there were those of conceptual perspective, logical simultaneity, perspectival incompatibility, intention, category, and relative acceptability. In what follows all of these notions will be treated once again, but this time with regard to the example of the present chapter, thus delineating more clearly their relation to actual science.

Of particular interest to the present view is the role played by *models* in the above example. Models will here take over the position occupied by *concepts* and *predicates* in Chapter 9. Thus while models and predicates, in application, may each be thought to constitute a *conceptual perspective*, scientific theories are conceived more particularly to be applied models.[7] It may be noted that theories are not here being identified with models: an unapplied model is not a theory;

[6] For details see e.g. Mitton (1939), pp. 179–182.

[7] In this regard cf. W. A. Wallace (1974), p. 263: "[I]n many situations where a novel modeling technique is employed to gain understanding of a phenomenon, a new way of looking at things is involved and a type of Gestalt switch may take place. In this sense Kuhn is quite correct in seeing scientific revolutions as involving such switches and changed viewpoints. In fact, his paradigm shifts can very frequently be seen as modeling shifts. ... "

and while every scientific theory is taken to be some sort of applied model, not every applied model need be a scientific theory.

On the present view, that to which a model is applied is determined by the *intention* of the individual applying it; and it may be applied either to real or to imaginary cases: e.g., the ideal gas model may be applied either to real or to ideal gases. We should thus say that the *reference* given the model, or the *intended domain* of the theory, differs in the two cases. More generally though, the intended domain of a scientific theory is to be thought of as encompassing all of the empirical states of affairs to which it is intended that its model be applied. Thus we see that the reference given van der Waals' model is broader than that given the ideal gas model, in that it is intended to be applied not only to substances in their purely gaseous form, but also to such substances when they undergo a change of state.

If the parameters a and b of the van der Waals equation of state are given the value zero, then van der Waals' equation reduces to the equation of state for an ideal gas. As a development of the present view, we can more generally say that the notion of *reduction* is to apply to those cases where a more complex equation gives results identical to those of a simpler equation when certain of the parameters occurring in the former are given limiting values (often zero). When this is so, we should thus say that the more complex equation *reduces to* the simpler one, or, we may say that the simpler equation, or the model from which it is derived, constitutes a *limiting case* of the more complex equation or model respectively. Now, it is to be noted that in the physical interpretation of van der Waals' equation the parameters a and b are not to be given the value zero, but are each rather to assume some constant positive value depending on the substance under investigation. Thus the reduction here is a purely formal one, and it does not obtain in any instance where van der Waals' equation is being treated as the expression of a law of nature, or as the equation of state of van der Waals' model.[8]

---

[8] The notion of reduction suggested here is essentially similar to the notion of *correspondence* treated e.g. in Krajewski (1977), Ch. 1: cf. esp. pp. 6 & 10. Note that the present notion is not intended to be the one employed when speaking in such contexts as that concerning e.g. the reduction of biology to physics.

It might also be noted at this point that the ideal gas model and that of van der Waals,

The fact that the parameters *a* and *b* in van der Waals' equation of state always receive some positive value, i.e., the fact that in his model inter-molecular forces are always assumed to be present and the molecules are always taken to have some non-negligible volume, means that his model and the ideal gas model cannot be applied simultaneously by one person to one and the same state of affairs. In present terms this is to say that they are *perspectivally incompatible* and cannot be treated *logically simultaneously* in application to the same intended domain. And this is so even if in their respective applications they were to suggest the same results.[9] But this is not to say that the two models contradict one another, or that they are in any sense incompatible when viewed as unapplied systems. First, they do not contradict each other since neither is a linguistic entity, let alone a proposition—i.e. neither model asserts anything to be the case. And second, while formal relations might be established between them, the most that such formal relations could be hoped to show is that they are different, not that they conflict. Viewed as unapplied systems, the models, in a sense, stand side by side, and neither impinges on the domain of the other. It is only in their application that conflict arises.

In the simpler case treated in the previous chapter, it was suggested that two predicates would conflict in application to one and the same thing only if both predicates were of the same *category*, or, in terms of the present chapter, only if they were different values of the same *parameter*. And in the present case we should say that two *models* would conflict in application to the same intended domain only if both models involve the same parameters. Furthermore, as in the previous chapter, evidence concerning whether or not two models do involve the same parameters may be afforded via a consideration of whether their actual application or testing requires the same (sorts of) *operations* or *measurements*. Thus the fact that the application of both the

---

in being conceptually distinct, strictly speaking constitute the respective bases of independent theories, though both fall under the more general heading of the kinetic theory of gases.

[9] Imagine, for example, Bernoulli's model and the ideal gas model to give identical results: they would nevertheless be perspectivally incompatible due to the assumption of an infinite number of molecules in the former and a finite number in the latter.

van der Waals model and the ideal gas model involves identical operations using a ruler (or a more sophisticated instrument), a manometer, and a thermometer suggests that both models involve the parameters volume, pressure, and temperature.

In the present case not only does the testing or application of the two models under consideration require the same sorts of operations involving the same sorts of instruments, but in certain applications the models suggest measurably different results of their common operations. When applied to gases at high pressures and low temperatures van der Waals' model, or its equation of state employed as the expression of a law, predicts values much closer to those obtained on measuring instruments than does the ideal gas model (Boyle's law, or the general gas law). Van der Waals' model is thus more *accurate* than the ideal gas model, and, other things being equal, we should consequently say that it is the more *acceptable* of the two, and thus that it constitutes a scientific *progression* beyond the other. This conclusion is further supported by the fact that the actual *scope* of van der Waals' model is much wider than that of the ideal gas model, for it suggests results that are at least of the correct order of magnitude in the case of change of state, while this case lies quite beyond what is tractable on the ideal gas model. On the other hand, the ideal gas model is *simpler* than that of van der Waals; this fact, plus the relatively adequate results obtained by Boyle's law in the case of gases at low pressures and high temperatures, prevents its being discarded from science.

6. THEORIES AS DISTINCT FROM LAWS

One of the major shortcomings of modern philosophy of science as a whole, and of the Empiricist and Popperian views in particular, has been its failure to provide a positive characterization of the essential difference between the natures of scientific laws and theories. On the present view however, as has been mentioned above, scientific theories are conceived to be applied models; and the expression of a (quantitative) scientific law, rather than take the form of an all-statement as on the Empiricist and Popperian views, is to consist of an *equation* relating certain *measurable parameters*. This conception of laws, and with it the role here seen to be played by parameters and measuring instruments, is in keeping with R. Fürth's characterization, where he says:

It is the generally accepted view [of physicists] that the laws of physics are expressed in the form of mathematical equations between certain variable quantities or 'parameters' which may either be capable of assuming any values within a certain range (continuous parameters) or are restricted to a finite or infinite, but enumerable set of discrete values (discontinuous parameters). A parameter can either be directly defined 'operationally', that is by a well defined process of measurement on a given physical system by means of certain measuring instruments, the readings on which determine the numerical values of the parameter; or it may be defined indirectly by a mathematical formula in terms of other directly defined parameters.[10]

On the present view, following the above example, scientific laws are seen to be of essentially two sorts, where one and the same law may qualify as being of both sorts. Thus, for example, Boyle's law may be arrived at solely on the basis of the performing of certain operations with measuring instruments, or, it may be arrived at via derivation from the ideal gas model. In the first case it is considered an *empirical* or *experimental* law,[11] and, in the second, it will here be termed a *theoretical* law.[12] Furthermore, it may be the case that

---

[10] Fürth (1969), p. 327; the whole of Fürth's paper constitutes a valuable discussion of the role of models in physics. In the present regard, cf. also Poincaré (1903), p. 217, and Campbell (1920), Ch. II.

[11] Cf. Campbell (1920), p. 153: "[W]hy do we call some laws "empirical" and associate with that term a slight element of distrust? Because such laws are not explained by any theory." An empirical law is here taken as not necessarily involving measurements—we might call such a law as does a *quantitative* law; and an experimental law is simply to be one which applies in experimental situations.

[12] This usage of the term differs from that of e.g. Carnap and Nagel, in which a theoretical law is necessarily to contain terms referring to unobservables. See e.g. Nagel (1961), p. 80, and Carnap (1966 *a*), p. 227. An interesting discussion of the issues raised in the present chapter may also be found in Hempel (1970). In these later writings, each of the above authors makes reference to Campbell, and their respective discussions are largely shaped by Campbell's (1920). Hempel in fact goes so far as to suggest that Campbell, who emphasizes so strongly the notion of analogy, is a proponent of what Hempel calls the 'standard conception' of scientific theories. The fact is that Campbell's view lies quite beyond the standard conception, if by this we understand Hempel to mean the logical empiricist conception, or anything closely resembling it. (Consider, e.g., Campbell's saying: "Of course the province and power of logic have

certain laws are only empirical, in that they have as yet not been successfully derived from any model or theory; or they may be only theoretical, in that they have been derived from some model, but, perhaps due to the absence of sufficiently sensitive measuring instruments, have not been confirmed by experiment. In this way then we should say that van der Waals' law is more a theoretical law than an empirical one, for, while it has received a certain amount of support from empirical investigations, it takes the particular form that it does as a result of the nature of the theory or model from which it has been derived, and empirical investigations alone would not have suggested its having exactly that form.

Scientific theories, on the other hand, while they may in certain cases be *expressed* by equations (relating the constituents of a model), are here positively characterized as being applied models—i.e. models from which empirical laws can be derived. Thus the present view is very much in keeping with that of N. R. Campbell where, in summary, he says:

> [T]he value of the [dynamic theory of gases] is derived largely, not from the formal constitution, but from an analogy displayed by the hypothesis.
>
> This analogy is essential to and inseparable from the theory and is not merely an aid to its formulation. Herein lies the difference between a law and a theory, a difference which is of the first importance.[13]

Now, in considering theories to be applied models, the present study is by no means suggesting that such models must be either picturable or mechanical; and part of what is to follow is based on the fact that they seldom, if ever, are the former.

Pierre Duhem, in the chapter of his book *The Aim and Structure of Physical Theory* entitled "Abstract Theories and Mechanical Mod-

---

been very greatly extended in recent years, but some of its essential features ... have remained unchanged; and any process of thought which does not show those features is still illogical. But illogical is not synonymous with erroneous. I believe that all important scientific thought is illogical, and that we shall be led into nothing but error if we try to force scientific reasoning into the forms prescribed by logical canons." (1920), p. 52.) What the above authors are actually doing in the works cited here is not so much providing elaborations of the Empiricist conception of theories as affording relatively neutral descriptions of the way theories function in science.

[13] Campbell (1920), p. 119; cf. also pp. 129–132.

els'', has argued against the necessity of employing mechanical models in scientific theorizing. To the extent that his remarks are taken to suggest that *no* models are required in theorizing, the present study must of course object. But a closer reading of his text suggests that what he is actually arguing against is the need of employing *picturable* models; and if this is so, there is no disagreement between his view and the present one.

Duhem begins his argument by distinguishing between the 'abstract mind' of the Continental physicist and the 'visualizing mind' of the English physicist; and he suggests that men possessing minds of the former sort ''have no difficulty in conceiving of an idea which abstraction has stripped of everything that would stimulate the sensual memory'',[14] while the intellectual power of the latter ''is subject to one condition; namely, the objects to which it is directed must be those falling within the purview of the senses, they must be tangible or visible.'' And at a later point he says, ''The French and German physicist conceives, in the space separating two conductors, abstract lines of force having no thickness or real existence; the English physicist materializes these lines and thickens them to the dimensions of a tube which he will fill with vulcanized rubber.''[15]

Now even if it is assumed that the ultimate aim of scientific theorizing *is* to produce a mechanical model of the sort of phenomenon under investigation, such a model need not be one of the sort which apparently concerns Duhem—i.e. it need not be a picturable one. Thus, for example, we might have a model employing abstractions like the idea of a mass-point—which cannot be pictured—while the model is nevertheless mechanical. (The same is true of the ideal gas model if we take its molecules as having no volume, rather than merely negligible volume.) But what is of importance here is that on the present conception scientific theories, while each is essentially related to some sort of model, need not be related to visualizable models, and as a matter of fact almost always involve abstractions or *idealizations* (like that of a mass-point) in their attempts to provide an understanding of the essence of certain natural phenomena.

---

[14] This quotation and the next are taken from Duhem (1906), p. 56.

[15] Duhem (1906), p. 70.

7. IDEALIZATION

In recent years a number of Polish philosophers have drawn attention to the importance of idealization in science, and to the fact that its existence runs counter to the conception of science suggested by the more traditional views. In this regard Leszek Nowak, for example, has said:

> Let us notice that the basic assumption of the inductivist model of science is the idea of science as a form of registration, or phenomena systematization. ...
>
> This idea is not beyond controversy, however. Let us take a look at the theoretical activities of a physicist. His basic activity is the construction of models of different sorts: the model of the perfectly rigid body, the model of the homogeneous cosmos, the model of the perfect gas, etc. The construction of such models always involves the omission of certain features of material objects .... Thus, where real physical objects have three spatial dimensions, plus mass, acceleration, etc., their model, the material point, has zero dimensions, and only certain (mass, acceleration, etc.) characteristics of the objects so modelled ... . Model construction is, then, anything but phenomena registration—the model is not "an abbreviated record" of a phenomenon but its "deformation".[16]

Nowak goes on to suggest that the viability of the model rests on whether the characteristics emphasized in it are *essential* to the phenomenon.[17]

In his book *Correspondence Principle and Growth of Science*, Władysław Krajewski has provided many examples involving idealization (including the ideal gas model and that of van der Waals) taken from both the physical and social sciences; and he points out how these sorts of cases differ from the 'crude empiricism' of Aristotle, in which idealization does not play a role.[18] Both Nowak and Krajewski have recognized that the employment of idealizations involves the making of *counter-factual* assumptions, and that it consequently creates serious problems for the Empiricist conception of scientific laws and theories. Nowak has also suggested that idealization is not in keeping with Popper's philosophy of science; and in the present study

---

[16] Nowak (1979), pp. 284–285.

[17] In this regard see also Hesse (1966), pp. 34f.

[18] Cf. Krajewski (1977), Ch. 2. For the present author's view on idealization, as expressed elsewhere, see Dilworth (1979), and Dilworth/Bunge (1979), p. 420.

the reason for this becomes clear, namely, that the Popperian conception, similarly to the Empiricist, is based on the Deductive Model (which, incidentally, closely resembles the Aristotelian syllogism, as has been noted in Chapter 1).

The present conception, on the other hand, recognizes the importance of idealization in science, and sees it as playing a central role in the determination of the natures of individual scientific theories. And, by taking scientific theories to be applied models, it is an easy step to the view that it is precisely such ideal entities as mass-points and perfectly elastic spheres, occurring in these models, that are the referents of theoretical terms.

8. THEORETICAL TERMS AND CORRESPONDENCE RULES

In Chapter 4 the notions of theoretical terms and correspondence rules were discussed in the context of the Empiricist conception of the relation between theories and laws, and were found to create grave problems for the Empiricist view.

Theoretical Terms

The Empiricist problem of theoretical terms concerns the *meaningfulness* of such terms, and results from the fact that on the Empiricist conception the meanings of all non-logical terms in science are ultimately to be determined on the basis of their formal relation to observational terms. And, in keeping with its Positivist beginnings, this view sees observational terms as being those the referents of which are entities of the sense-data variety. As suggested in Chapter 4, this requirement creates problems not only with regard to terms of the sort mentioned in the previous section, but also with regard to mensural terms, and terms the referents of which have dispositional properties.[19] Unfortunately, due to the similar intractability of terms of this latter sort on the Empiricist conception, they have also come to be called theoretical, thus confusing the distinction between empirical or experimental laws on the one hand, and scientific theories on the other.

[19] For a discussion of dispositional properties which is in keeping with the view of the present study, see Agazzi (1976), pp. 149 ff.

8–812581 Dilworth

On the present conception, however, science is not seen to rest ultimately on that which can be directly experienced—subjective sense-data, but on the results of operations involving the employment of certain instruments—results which can be intersubjectively shared.[20] Thus mensural notions like that of temperature are not viewed as being theoretical (which is not to say that they are free of a priori presuppositions), but as being empirical or experimental, since they are employed in the expression of empirical laws which, taken together, may be seen to constitute the whole of the empirical basis of science.

The present view also differs in important ways from that of P. W. Bridgman, though he too emphasizes the role played by operations in science. The major difference between his view and the present one lies in the stress he places on the *meanings* of scientific terms, and with it his suggestion that their meanings are ultimately determined by the operations performed in their application.[21] Here, on the other hand, meanings do not play a central role, but rather the operations do so themselves. Thus, as has been suggested earlier, two different theories employing terms with quite different meanings may both be related to the same operations. But of greater importance at this point is that on the present view theoretical notions are not seen to depend on operations in order to be meaningful, and it is in fact often the case that an antecedent appreciation of their 'meanings' suggests which operations may be performed to determine their relative applicability. Thus the present view makes a clear departure from both Bridgman and the Empiricists by suggesting that theoretical terms seldom, if ever, obtain their meanings from the observational basis to which the theory is intended to apply,[22] but obtain them rather from any source whatever—sources as varied as metaphysics and everyday discourse.

Thus, in the case of the kinetic theory of gases, we see that the basic theoretical notions are obtained from the metaphysics of atom-

---

[20] With regard to the relation between the performing of operations and intersubjectivity in science, see Agazzi (1977*a*), pp. 162 ff., and (1978), pp. 100 ff.

[21] Cf. e.g. Bridgman (1936), Chs. II & IV.

[22] For a lucid critique which is of relevance to this and other points raised in the present chapter, see Spector (1965). Concerning the present point, see also Kuhn (1974), pp. 465–466.

ism (though they would have been just as 'meaningful' if newly introduced), and from such everyday concepts as random motion and elasticity. And while the molecules in an ideal gas are to move in a way which is in keeping with Newton's laws of motion, the meanings of the terms used in describing the model can hardly as a consequence be said to be dependent on those laws, much less on the fact that the laws might themselves be considered empirical.

Correspondence Rules

As mentioned in Chapter 4, the Empiricist problem of how theoretical terms obtain their meanings is closely linked to their problem of how empirical laws can be formally derived from theories containing such terms. And this latter problem also arises for the Popperian view, for, in being based on the Deductive Model, it too requires a formal derivation from the theoretical level to the empirical. In recognizing the impossibility of a direct derivation, Empiricists have noted that in actual science certain 'rules' (correspondence rules) are required in order to connect theoretical notions to empirical ones. (The question of whether, given these rules, the resulting derivation is a strictly formal one will here be left open.) Thus, for example, Hempel has said:

> In the classical kinetic theory of gases, the internal principles are assumptions about the gas molecules; they concern their size, their mass, their large number; and they include also various laws, partly taken over from classical mechanics, partly statistical in nature, pertaining to the motions and collisions of the molecules, and to the resulting changes in their momenta and energies. The bridge principles [correspondence rules] include statements such as that the temperature of a gas is proportional to the mean kinetic energy of its molecules, and that the rates at which different gases diffuse through the walls of a container are proportional to the numbers of molecules of the gases in question and to their average speeds.[23]

On the present view however—where models are taken to play a central role in theorizing—rather than posing a problem, correspondence rules are seen to be precisely what is required in order for the analogy displayed by the model to be applicable to empirical situations. Thus the present view accepts the existence of correspondence

[23] Hempel (1970), p. 144. See also e.g. Nagel (1961), pp. 93–94.

rules, and in fact requires them in order for a theory to be applicable. In contrast with this, the admission of their existence on the part of Empiricists is at the same time an admission that empirical laws cannot be derived from abstract theories without them—i.e. that the derivation is not direct, but indirect, since it depends on these rules. That the derivation is indirect is made all the clearer when we realize that the correspondence rules are often only tacitly given, and that their nature is not dictated by the model (theory) itself. As regards this latter point, we see that such rules are more like conventions determined by scientists in order that the theory *be* applicable to certain empirical states of affairs. Thus we also see that one and the same model, when applied to different sorts of situations, will have different correspondence rules. On the present view we should thus say that the correspondence rules employed in the case of applying a certain model are determined partly by the nature of the model itself, and partly by the *intention* of the person(s) applying it. And to this we may add that the application of the model is to consist in the derivation of its empirically testable theoretical laws with the aid of just such rules.

9. REALISM VS. INSTRUMENTALISM

On the present view the debate between realism and instrumentalism with regard to the ontological status of theoretical entities is left open. It may be pointed out however that the present view suggests that to the extent that theoretical notions are idealizations or abstractions, their real counterparts, should they exist, cannot have exactly those properties supposed by the notions. For example, granting that real gases are composed of molecules in rapid motion, these molecules could not have zero volume (as one interpretation of the ideal gas model might suggest), nor, it seems, could they be perfectly spherical or perfectly elastic.[24]

On the other hand, to deny that they exist at all might also appear

---

[24] In this regard cf. Boltzmann (1896), p. 26, where he says: "In describing the theory of gases as a mechanical *analogy*, we have already indicated, by the choice of this word, how far removed we are from that viewpoint which would see in visible matter the true properties of the smallest particles of the body."

problematic given that we understand theories as providing explanations either of empirical laws or of individual phenomena. A natural interpretation of the role of theories and laws suggests that while an empirical law taken by itself only evinces a form of constant conjunction, in certain cases when derived from a theory it might attain the status of a *causal* law. For example, having derived Boyle's law from the ideal gas model we might say that the pressure of a gas increases with decreasing volume *because* the number of molecules striking the walls of its container is thereby increased per unit time, and that in this way we have consequently explained the phenomenon. We have explained that which at first we did not understand: the constant conjunction, in terms of that which we do: the model.[25] And it might be though that to accept this view requires some sort of commitment to the existence of molecules in real gases.

Now, while the above conclusion might itself be challenged, what is being stressed here is simply that the present view excludes neither a realist nor an instrumentalist approach to the nature of scientific theories.

The development of the present conception given in this chapter should not only have helped clarify its connection to actual science, but it should also have revealed further difficulties faced by similar attempts to apply the Empiricist and Popperian conceptions. Of major importance in this regard is the latters' consideration of theories as being universal propositions which are formally to entail the descriptions of certain sorts of phenomena, whereas in actual fact, as is suggested by the present view, theories are more like models which bear, not a logical relation to (the descriptions of) phenomena, but an analogical relation to them. In the next chapter an alternative account to the present one will be considered, in which a notion of model is also to play a fundamental role.

---

[25] We would thus have an instance of what Campbell calls 'explanation by greater familiarity' where he suggests that "The theory of gases explains Boyle's Law, not only because it shows that it can be regarded as a consequence of the same general principle as Gay-Lussac's Law, but also because it associates both laws with the more familiar ideas of the motion of elastic particles." (1920), p. 146.

With regard to the idea of models characterizing the essence of phenomena, and thereby causally explaining them, see also Nowak (1980), pp. 128–129.

CHAPTER 11

THE SET-THEORETIC CONCEPTION OF SCIENCE

1. A NEW FORMAL APPROACH TO SCIENCE

In recent years a new view has emerged in the philosophy of science, taking as its basis the informal axiomatization of Newtonian particle mechanics in terms of a set theoretical predicate. This axiomatization itself appears first in McKinsey, Sugar, and Suppes (1953), and, in an attempt to handle theory dynamics, has been further developed by J. D. Sneed in his book *The Logical Structure of Mathematical Physics* (1971). In this book Sneed attempts to reconstruct Newtonian particle mechanics in such a way as to clarify the role of theoretical terms in science, and to provide a conception of how scientific theories can rationally evolve in the face of recalcitrant data.

The above approach, constituting the set-theoretic or 'structuralist' conception, is similar to the one of the present study to the extent that it does not see scientific theories as simply being universal statements. On the other hand however, it is in a more fundamental respect similar to the Empiricist and Popperian views, for its basis is a particular *formal* system, namely, intuitive set theory.

In the present chapter a critique will be made of the set-theoretic view as it is presented by Sneed, and by Wolfgang Stegmüller.[1] A key feature distinguishing Sneed's approach from that of his predecessors is the introduction and extensive treatment of what Sneed calls 'the problem of theoretical terms'.

2. SNEED'S PROBLEM OF THEORETICAL TERMS

As discussed in Chapters 4 and 10 above, the Empiricist problem of theoretical terms has its origin in the logical positivist criterion of empirical meaningfulness: that a non-formal proposition is meaningful

---

[1] Cf. Stegmüller (1973).

only if it is verifiable by direct observation. And, as has been pointed out in those places, this problem arises not only in the case of terms like "mass-point" and "electron", but also in the case of such mensural terms as "mass" and "temperature", as well as in the case of terms like "magnet", whose referents have dispositional properties; and it has led to the classification of these latter sorts of terms as also being theoretical. On the Empiricist conception of science then, if a solution to this problem of 'theoretical' terms were to be found, it would have to lie in the direction of reducing such terms to terms the referents of which are directly observable.

Sneed's problem of theoretical terms, on the other hand, is quite different, for it does not concern the Empiricist conception, but the set-theoretic one; and, of the three sorts of terms mentioned above, it pertains only to those of the mensural variety.

Sneed's Notion of Theoretical Term

For Sneed, a theoretical term is a function any measurement of the value of which implies that the theory in which the function occurs has already been successfully applied.[2] Thus the question as to whether a particular function is theoretical must be relativized to a particular theory. Where a function is theoretical relative to a theory θ, it is called θ-dependent. Sneed takes the mass and force functions of Newtonian particle mechanics to be functions of this kind, while, for example, the position function is not. In evaluating his concept of theoretical term we might thus consider the example he gives as his main support for treating "mass" as theoretical relative to classical particle mechanics:

> An example of a θ-dependent function is the mass function in an application of classical particle mechanics to a projectile problem. In this case we typically determine the mass of the projectile by "comparing" it to some standard body with a device like an analytical balance or an Atwood's machine. ... But the *only* reason we believe that these comparison procedures yield mass-ratios, and not just numbers, completely unrelated to classical particle mechanics, is that we believe classical particle mechanics applies (at least approximately) to the physical systems used to make the comparisons. If someone asks why the number $(a/g-1\ /\ a/g+1)$, calculated from the

---

[2] Cf. Sneed (1971), pp. 31 ff. & 116 f.

acceleration observed in an Atwood's machine experiment, is the mass-ratio of the two bodies involved, we reply by deriving it from the application of classical particle mechanics to this system. I maintain that examination of *any* acceptable account of how the mass of a projectile might be determined would reveal the same sort of dependence on an assumption that classical particle mechanics applied to the physical system used in making the mass determination.[3]

To begin, we might ask to what extent this example supports Sneed's claim that "mass" is theoretical in his sense. While it may here be granted that the use of an Atwood machine to determine relative masses or 'mass-ratios' does presuppose the viability of Newton's second law, we note that what Sneed has described is not actually an experiment (test), nor is it a determination of the sort the machine is usually employed to make (which is that involving the establishment of g); what is more, both this latter use and that to which the machine was originally put (the results of which should allow it to be employed in the way described by Sneed) presuppose that the mass-ratios in question are already known. Thus, with regard to its original use for example, we read in Hanson: "... Atwood showed that when $m_1=48$ gm. and $m_2=50$ gm. then the acceleration of m_2 is indeed 20 cm./sec.^{2}. The results were carefully recorded and generalized: they squared with the predictions of the second law. For Atwood this fully confirmed the law."[4] Taking Hanson's description to be correct, how then did Atwood determine the values for m_1 and m_2? He could not have used his machine to do so, on pain of circularity. He most probably measured these masses directly, using a beam balance, or some similar device. Thus, to say the least, Sneed has chosen a rather peculiar example to support his claim that *every* measurement of mass presupposes that classical particle mechanics applies to the system used in making the mass determination. Why did he not treat instead the more straightforward case involving the use of a beam balance? One reason is perhaps that in this case he might simply have found that mass can be and was determined without presupposing that particle mechanics applies, since this sort

---

[3] Sneed (1971), pp. 32–33. (We take this occasion to remark that an Atwood machine is essentially a pulley over which is passed a thread whose ends are connected to weights of unequal mass.)

[4] Hanson (1958), p. 101.

of balance has been in use since long before the time of Newton. At any event, until Sneed explains how the determination of the mass of some thing by the use of a device like a common beam balance presupposes particle mechanics to apply, he is hardly justified in claiming "mass" to be theoretical in his sense.

Noting this then, and that the same sort of comment applies to Sneed's treatment of force—since he does little more than simply claim its measurement to presuppose particle mechanics, we might go on to consider whether, even if he had shown "mass" and "force" to be theoretical in his sense, his sense is one which can be generally accepted as capturing the essence of what theoretical terms actually are. In this regard we note first that Sneed does not seem to see scientific laws as being anything other than the constituents of scientific theories, and more often than not in discussing the referents of what he considers to be theoretical terms he speaks of their measurement presupposing some law. Thus, for example, in giving his reason for considering "force" to be theoretical, he says: "All means of measuring forces, known to me, appear to rest, in a quite straightforward way, on the assumption that Newton's second law is true in some physical system, and indeed also on the assumption that some particular force law holds."[5] However, while it is not being claimed in the present study that Newton's second law is a straightforwardly empirical one, all the same it is one thing to say that certain sorts of measurements depend on a particular *theory* (classical particle mechanics), and quite another to say that they rest on a particular *law*, even if the law is intimately related to the theory. And it seems strange to call e.g. "force" theoretical because its measurement should presuppose some law. This point concerns more than mere terminology, for there are views of science in which measurements are based on laws, not on theories, and in which, for this reason, no mensural term is to be considered theoretical.[6]

More generally, when we consider the extent to which Sneed's distinction between what he calls theoretical and non-theoretical

---

[5] Sneed (1971), p. 117.

[6] Cf. e.g. Campbell (1928), p. 1: "[T]rue measurement, as we shall see ... is based upon ... scientific laws. We shall be nearer the truth if we define measurement as the assignment of numerals to present properties *in accordance with scientific laws*."

terms serves to clarify this distinction as it occurs in actual science, we see that not only are Sneed's theoretical terms mensural terms, but his distinction provides no positive understanding of what an empirical term should be. And to this it must be added that his notion of theoretical term does not pertain at all to those sorts of terms which are normally considered to be particularly exemplary of theoretical terms, e.g. terms like "mass-point" and "electron", the referents of which are unobservable.

The present critique of Sneed's notion of theoretical term is not intended to suggest, however, that the notions of mass and force in Newtonian mechanics are unproblematic, for there of course exist the age-old problems of providing an intuitively satisfying definition of "mass", and of deciding whether Newton's second law ought to be considered as anything more than merely a definition of "force". Nor is it the present intention to suggest that in testing a particular theory, or a law, there do not occur situations in which the theory or law being tested must in some sense, or to some extent, itself be presupposed in order to carry out the test.[7] But the very existence of these sorts of cases draws attention to a deeper issue regarding the viability of Sneed's notion of a theoretical term, namely, that he introduces and almost always employs the notion in the context of his own conception of science, and it is this use of it which he ultimately must justify. Thus, for example, even if he were to succeed in showing that in actual science any measurement of mass in some sense presupposes the applicability of classical particle mechanics, he would still have to show that the essence of this phenomenon is retained when we move over to thinking of 'mass' as a function occurring in the definition (or extension) of a set theoretical predicate, and the successful application of a theory as consisting in the discovery of a set theoretical model for such a predicate.

Sneed's Problem of Theoretical Terms

Sneed presents his 'problem of theoretical terms' as a problem which can arise, given his notion of theoretical term, for a relatively

---

[7] For a stimulating discussion of this problem (in Swedish) with regard to Ohm's law, see Johansson (1978); concerning the conventional aspects of Newton's laws of motion, see Ellis (1965).

straightforward construal of his own view of science. On Sneed's view, i.e. on the set-theoretic conception, a *theory* (of mathematical physics) is to consist of a 'formal mathematical structure' characterized by a set theoretical predicate, plus a set of physical systems each of which is an 'intended application' of the theory.[8] An *empirical claim* of the theory is a statement to the effect that the predicate is in fact applicable to one of its intended applications, i.e., that one of these physical systems is a set theoretical model of the predicate. And for Sneed the problem of theoretical terms is one of providing a conception of how the truth value of such a claim may be determined given that the definition of the predicate might well contain terms which are theoretical in his own sense.

Sneed takes the possible existence of this problem for the set-theoretic view as being sufficient motivation for the rest of his discussion, and, presupposing this view, devotes a major part of his book to reasonings in terms of his theoretical/non-theoretical distinction. His efforts in attempting to solve his problem involve modifications on the method (discussed in Chapter 4) taken over by the Empiricists from Ramsey in an attempt to solve their problem of theoretical terms. Unfortunately, however, in spite of his lengthy development of 'Ramsey eliminability', the results he obtains seem to apply in a straightforward way only when the values of the 'theoretical' functions occurring in the 'formal mathematical structure' are in principle determinable by computation from the values of the 'non-theoretical' functions; and, in the case of particle mechanics, Sneed in the end concludes that "it is not likely that the mass function can be Ramsey eliminated from *any* claim of classical particle mechanics."[9]

---

[8] Cf. Sneed (1971), pp. 36 & 161. In its simplest form a theory is to be, more particularly, an ordered pair, the first element of which is the set of models of the predicate, and the second element of which is the set of those empirical states of affairs which it is intended should be models of the predicate: cf. e.g. Adams (1959), p. 259. Though in his book Sneed makes of the first element an ordered *n*-tuple containing sets of possible models, possible partial models, constraints and so on, the essence of the approach outlined in Adams (1959) is retained. Both elements in the set-theoretic notion of a theory run counter to one's intuitions: the first because on Sneed's reasoning *any* model—physical or otherwise—of the predicate should constitute a part of the theory's formal mathematical structure, and the second because those states of affairs to which it is intended that the theory be applied are conceived as being part of the theory itself.

[9] Sneed (1971), p. 152.

In any case, whether Sneed has or has not overcome his problem of theoretical terms is of little consequence to the present critique, for his problem is couched in the context of his own conception of science, and it is this conception which is here being called into question. But, from the present point of view, what is of importance here is the fact that both Sneed's theoretical/non-theoretical distinction as employed in the context of the set-theoretic conception, and his problem of theoretical terms, assume as a fundamental presupposition that the models of set theoretical predicates characterizing theories of mathematical physics may be found in the real world—that it is through the determination of just such models that these theories are to apply to empirical reality. As the present chapter proceeds, however, the attempt will be made to show that this idea is quite mistaken.

As mentioned above, on the set-theoretic view a scientific theory is to consist of a certain 'mathematical structure' and a set of 'intended applications', the latter being the set of real states of affairs to which it is intended that the theory apply. But, as will now be shown, one indication of the incorrectness of thinking of the models of predicates characterizing scientific theories as existing in the real world lies in the inability of the set-theoretic conception to delineate this set of 'intended applications' so long as it restricts itself to employing the methods afforded by set theory.

3. THE PROBLEM OF DELINEATING THE INTENDED DOMAIN

In attempting to identify the realm of intended applications of a theory on the set-theoretic view, we might begin by considering as a candidate the set of (set theoretical) models of the predicate used to characterize the mathematical structure of the theory. Were we to do this however, we would quickly see that the adoption of this set of models, i.e. this set of sets to which the predicate correctly applies—the predicate's *extension*, would leave us saying that the theory is intended to apply to all and only those states of affairs to which it in fact does apply; and this of course would not at all capture the notion we are after.[10] In an attempt to overcome this difficulty, Sneed conceives of those states of affairs to which it is intended that a theory be applied as being a subset of the set of models to which a sub-part of the predicate applies. Of course the theory's extension is itself such a subset, and so the theory's set of 'intended applications' requires further distinguishing characterization.

At this point Sneed (as well as Stegmüller, though his treatment of the case is slightly different)[11] quite transcends the methods afforded by set theory in filling out his picture of science. Using the notation of the set-theoretic conception, we have at present that the set I of intended applications of a theory is a subset of the set M_{pp} of models of a particular part of the predicate characterizing the theory's mathematical structure (this part of the predicate is not to contain definitions of 'theoretical' functions, and its models are thus to be 'empirically' recognizable as such); i.e., we have that $I \subseteq M_{pp}$, where M_{pp} is the set of models of the restricted predicate. We note that we also have $M \subseteq M_{pp}$, where we here take M to be the set of models of the complete predicate. And the question is then: how is I to be characterized so as to distinguish it from M?[12]

Sneed here turns to the *beliefs* of individual persons, e.g. physicists, and suggests that there exists for them a set of *paradigm examples* (*à la* Wittgenstein) $I_0 \subseteq I$ each member of which is believed to be extendible so as to constitute an element of M, and which furthermore characterizes I in that all members of I bear a relation of the same sort as Wittgenstein's 'family resemblance' to the members of I_0.[13] I is further characterized by requiring that it meet certain other conditions, one of which is that its members be physical systems (though Sneed is unable to tell us what is to count as a physical

---

[10] For a discussion of this point and others relevant to the present critique, see Hempel (1970), pp. 150 ff.; with regard to the problem of delineating the intended domain see also Agazzi (1976), pp. 144–145, and Nagel (1961), pp. 93 ff.

[11] Cf. Stegmüller (1973), pp. 193 ff.

[12] With regard to this question we note that there exist set theoretical models of e.g. classical particle mechanics to which it would normally never be intended that the theory be applied. Cf. e.g. Adams (1959), pp. 258–259, where the set P of 'particles' in a model of the theory has the number 1 as its sole member.

[13] Cf. Sneed (1971), pp. 268–270.

system on his view);[14] and instances of paradigm examples —members of I_0—in the case of classical particle mechanics are suggested to be the solar system and its various subsystems.[15]

The point to be made here is similar to those raised in Chapter 6 with regard to Popper's and Lakatos' transcendence of the Deductive Model. Irrespective of the extent to which the above description does or does not capture the way in which the intended domain of application of a theory is determined by physicists, by referring e.g. to their (potentially vacillating) beliefs it reaches beyond what is warranted on the basis of a conception of scientific theories in which they are formulated in terms of set theoretical predicates, and in this way is quite ad hoc.[16] Not only this, but such a notion as that of characterizing a set by reference to some vague trait similar to that of 'family resemblance' in fact runs counter to the spirit of intuitive set theory, on which Sneed ultimately bases his reasoning.

Thus we see that the set-theoretic approach has been unable to provide a coherent conception capable of accounting for the way in which scientific theories are related to those states of affairs to which they are *intended* to apply. And, as will be seen below, the concepts afforded by set theory function no better when it comes to handling the nature of the relation between scientific theories and those real states of affairs to which they are thought to *succeed* in applying.

4. THE PROBLEM OF EXTENSION

In its simplest form, the problem faced by the set-theoretic conception with regard to extension is that there *are* no physical models which satisfy the mathematical structure of a scientific theory axiomatized by a set theoretical predicate. For example, the solar system is not, nor is it presently believed to be, a model of Newtonian particle mechanics. Evidence for its not being a model is, e.g., the inability of

---

[14] Sneed (1971), p. 250.

[15] Stegmüller (1973), p. 175.

[16] A similar charge can also be levelled both at Sneed's definition of 'having a theory' (1971, p. 266), which presupposes such indistinct notions as 'a person having observational evidence', and at Stegmüller's development of Sneed's views in 'pragmatic' or 'extra-logical' terms: cf. e.g. Stegmüller (1976), p. 154, and (1973), p. 162.

Newtonian theory to account for the rate of the advance of the perihelion of Mercury. And, more generally, even in cases where theories are thought to apply quite well, they do not apply exactly, as they must in order for that to which they are being applied to constitute their model in the set theoretical sense.[17]

But an even stronger claim can be made here: it is that no physical system *could* be a set theoretical model of any modern scientific theory. And the reason for this is that such theories, by their very nature, involve *idealizations* (as discussed in the previous chapter) which could not exist in empirical reality. Sneed himself realizes the existence of this problem for his view:

> It might be argued that one simply cannot expect to find "real" physical models for *S*, or anything like it. If we limit ourselves to domains consisting of physical objects or other non-abstract individuals (e.g. sounds or events) which we can, in some way, perceive, and define functions in terms of "observable" relations among these objects, then ['*Q* is an *S*'] will always (or almost always) be false. Speaking platonistically, *S* is an ideal of which real objects are, at best, imperfect copies.[18]

Sneed then sketches two ways this problem might be approached—each of which would require a treatment transcending what could be provided by the concepts of orthodox set theory—and admits that he does not see how either way is to be worked out in detail. But he then suggests that the questions of logical structure he intends to examine—namely, those concerning the use of set theoretical predicates characterizing scientific theories to make empirical claims—can be considered independently of this matter. But can they? If the set theoretical predicates representing scientific theories can in principle only have abstract models, then no state of affairs of which Sneed's 'empirical claims' may be true could be an empirical one. And all of

---

[17] The attempt has been made in Moulines (1976) to overcome this sort of problem via the introduction of the notion of a 'fuzzy set'. But even aside from the fact that this move is not at all indicated by the set theoretical basis Moulines assumes, it involves the counter-intuitive assumption that scientific theories themselves are inexact in their predictions, rather than that their predictions, while exact, are not identical with the results obtained in the performance of actual measurements.

[18] Sneed (1971), p. 24; cf. also Sneed's remarks on pp. 112f. & 120. Note that the point being raised here concerns those sorts of entities which could not exist in the real world, not those which, while they might exist, would be impossible to detect.

Sneed's reasoning concerning the role of what he calls theoretical terms in the making of what he calls empirical claims would have no bearing on the question of the nature of the relation between scientific theories and empirical states of affairs. It is here suggested that this is in fact the case: that the models of modern scientific theories characterized by set theoretical predicates are such that they could not exist in the empirical world, and that as a consequence the major portion of Sneed's work bears little relation to actual science, since it is based on the belief that empirical states of affairs might, at least in principle, constitute such models.

The fact that the set theoretical models of predicates used to characterize scientific theories themselves involve idealizations suggests rather that such 'models' may be thought to be of the same sort as the (ideal) models treated in the previous chapter. In this case one might thus apply oneself to characterizing such ideal models in science in terms of set theoretical predicates. But it may be noted that such an exercise in formal reconstruction would at most succeed in presenting these models in a new form, and could in itself be expected to contribute little toward our understanding of the role that they play in science.

The considerations of the last two sections reveal then that the employment of the tools of orthodox set theory is not sufficient to capture either the notion of that to which a theory is intended to apply, or the notion of that to which it is thought to succeed in applying. But even if we set these failings aside, we find that the set-theoretic conception faces yet another problem: that of providing an account of progressive theory succession.

5. THE PROBLEM OF PROGRESS

The central notion in the set-theoretic view of how one theory might be conceived to be superior to another is that of *reduction*, and the idea it is intended to capture is, similar to the Empiricist notion of progress treated in Chapter 4, that a superior reducing theory T' should be able to explain all that an inferior reduced theory T is capable of explaining, and more besides.[19] The reduction relation itself is essentially conceived to be a one-many relation (i.e. the converse of a function) from a subset of the set M_{pp} of partial possible models of the reduced theory T into the set M'_{pp} of partial possible models of the reducing theory T'. More particularly, it should be a relation from the set I of intended applications of T to the set I' of intended applications of T'. Furthermore, the models x and x' (elements of M_{pp} and M'_{pp} respectively) which correspond by virtue of this one-many relation are, in some sense, to be identical. According to Sneed, the fact that to each x there may correspond more than one x' suggests that T' is more complete, precise, and provides the means for making more distinctions than does T. Also, the reduction relation is to be such that if x' is a model of the predicate characterizing T', then x is a model of the predicate characterizing T.

One example where the sort of reduction sketched above is to obtain involves rigid body mechanics (T) and particle mechanics (T'), when each is defined as a set theoretical predicate. Thus, for example, as regards the above requirement concerning identity, Sneed suggests that "it is possible to regard a rigid body and the particles that compose it as being, in some sense, the same thing."[20] But if we consider the relevant axiomatizations from which Sneed is working, namely, those given by Ernest Adams, we see that a rigid body and a particle are not (partial possible) models of their respective theories, but are members of members of such models. It is to be granted, however, that since on Adams' axiomatization the first axiom of rigid body mechanics states that the first five (of seven) elements of a system (ordered n-tuple) of rigid body mechanics themselves constitute a system of particle mechanics, every model of the former theory is a model of the latter. In this case then we do have a relation of identity, though it is not one-many, but one-one, and does not involve intended applications, but models.

---

[19] For the notion of reduction given in the present paragraph, see Adams (1959), pp. 256ff., Sneed (1971), pp. 216ff., (1976), pp. 135ff., and Stegmüller (1973), pp. 127ff. Though Sneed extends Adams' notion in order for it to cover his own more complicated concept of a theory, what is presented here should capture its essence when employed both by Sneed and by Stegmüller.

[20] Sneed (1971), p. 219.

Problems arise at this point, however, even if we simply follow the above line of reasoning, for it suggests that the set of models of a reduced theory is to constitute a subset of that of the reducing theory (and thus we might have a one-many relation between these sets). But this sort of relation would obtain in any case where the set of models of a predicate defined by certain axioms is compared with the set of models of a predicate whose definition is given by some but not all of those same axioms. Here the axioms of the reducing theory would be derivable from those of the reduced theory—not vice versa;[21] and 'progress' would result simply from the creation of new theories having fewer axioms than the old. Furthermore, far from being the more precise, as Sneed suggests, in this case the 'reducing' theory would be the less precise, in that fewer conditions would have to be met in order for a state of affairs to constitute one of its models.

However, as mentioned above, the reduction relation is not necessarily to obtain between M and M', but between I and I', and it is to be such that if x' is an element of M', then the x which corresponds to it by virtue of (the converse of) the relation is to be an element of M. Of course the conceiving of the reduction relation as obtaining between sets of intended applications is going to mean that it cannot receive a satisfying formal treatment until the problem of delineating the intended domain is solved. But even if (following Stegmüller) we were to conceive of the relation as being from one set of partial possible models onto another, it still faces the obvious difficulty that neither it nor its converse has been defined.[22] In other words, given some partial possible model of some theory, we are not told how to determine whether it corresponds to some partial possible model of some other theory, and so in no case can we be sure that one theory reduces to another in the set-theoretic sense. What we might do though is take the fulfillment of the conditions given at the beginning of this section as being *sufficient* for saying that one theory reduces to another, in which case (setting aside Sneed's identity requirement) an explicit definition of the relation between the sets of partial possible models would not be necessary. But, as has been noted by others,

---

[21] For a similar point, see Mayr (1976), pp. 279 ff.

[22] This point has also been raised in Kaiser (1979), p. 16.

this would allow the construction of a 'reducing' relation between almost any two theories.[23]

In a more general vein, even if the set-theoretic conception of reduction were able to accomplish all that it set out to accomplish, it would provide no conception of how successive theories may be incommensurable. We might say that two such theories were incommensurable if, when defined as set theoretical predicates, they had quite different models—i.e. if their respective models were in no sense the same; but then neither would be reducible to the other. And if it were concluded from this that successive theories are then not incommensurable, it might still be asked how it is that they often conflict with one another. If we look at the situation simply from the point of view of set theory, each of the theories should have its own set of models, and it should or should not be the case that one of these sets is a subset of the other, or that the two sets overlap. But it is not possible to construe the two theories as conflicting, for even if the axioms defining one of them were to include explicit negations of axioms defining the other, this would only mean that the theories had different models, i.e. that they would apply, if at all, to different things.

6. CONCLUDING REMARKS

From the point of view of the present study the axiomatization of Newtonian particle mechanics via the definition of a set theoretical predicate is in itself quite acceptable. At the same time however it is noted that there are other, equally acceptable, ways of expressing the definitions fundamental to classical particle mechanics. But what is here seen to be a basic problem for the set-theoretic view as expounded by Sneed and Stegmüller is its assumption that the models arrived at via the principle of abstraction from the axioms in the definition are to exist in the real world—that they are or can be that to which the theory is intended ultimately to apply. Here these models are seen rather to be abstractions of the same sort as is the geometrical space defined by the Euclidean postulates, or the ideal gas models as de-

[23] E.g. on p. 262 of his (1959) Adams himself makes a remark to this effect.

fined in the previous chapter. This view of the situation explains why "the distinction between [physics and pure mathematics] is not clear from a set-theoretical standpoint",[24] and also why this approach faces problems whenever questions regarding the empirical aspects of science are raised.

Thus where the set-theoretic conception is unable to explain such features of science as the intended domain of application of a theory, and how a theory may be considered successful though it does not quite fit the data, the present conception, with its basis in the Gestalt Model, is able to do so. Though Thomas Kuhn has himself suggested that the approach of Sneed and Stegmüller succeeds in capturing important aspects of his own view of science, the above considerations indicate that in the case of progressive theory succession —with which the present study is primarily concerned—the set-theoretic notion of reduction is unable to account either for theory conflict or for incommensurability, while the present conception provides an explanation of both of these aspects of science.

In the next and concluding chapter a brief look will again be taken at Newtonian mechanics, as well as at the laws of motion of Kepler and Galileo, and a sketch will be given of the approach to their views suggested by the present study. The opportunity will also be taken in this context to compare the present approach with the alternatives treated above on those points where the differences are particularly salient.

---

[24] Suppes (1967), Ch. 2, p. 52.

CHAPTER 12

APPLICATION OF THE PERSPECTIVIST CONCEPTION OF THE VIEWS OF NEWTON KEPLER AND GALILEO

1. NEWTON'S THEORY OF GRAVITATION

What is commonly termed 'Newton's theory of gravitation' is presented in his book *Mathematical Principles of Natural Philosophy* (1687) in the following way. Eight definitions are first given, including, for example, the definition of quantity of matter (mass) as the product of density and volume. These definitions afforded, Newton next presents three *axioms*, or *laws of motion*, which are:

1. Every body continues in its state of rest, or of uniform motion in a straight line, unless it is compelled to change that state by forces impressed on it.
2. Change of motion is proportional to the force impressed, and is made in the direction of the straight line in which the force is impressed.
3. The forces two bodies exert on each other are always equal and opposite in direction.

From these axioms six corollaries are derived which include, for example, an explanation of how a body should move when acted upon by two forces simultaneously. Following this begins Book I: The Motion of Bodies, and the first pages of this part of the treatise are devoted to a development of the infinitesimal calculus (in a geometrical form), which later serves as an aid to the determination of, for example, centripetal forces—i.e. forces directed to a single point.

From this basis then, Newton proceeds to expound his conception of the nature of the motion of bodies, often invoking as a working hypothesis that which, when generalized, has come to be called his *law of gravitation*. This law may be expressed as follows:

Any two bodies attract each other with a force which is proportional to the product of their masses and in inverse ratio to the square of the distance between them.

Though Newton does not himself present this law in his book, it is clear from his assumption of it in particular cases that it underlies the whole of his thinking with regard to the motion of bodies, and might just as well have been included as a fourth axiom at the basis of his system. In any case, for the sake of simplicity we may here say that Newton's three axioms or laws of motion, together with his law of gravitation, define the basis of what is commonly called Newton's *theory of gravitation*.

The question often arises in considerations of Newton's theory as to whether his axioms or laws of motion ought best to be taken as definitions (or a priori truths), or as empirical laws, and the conclusion usually drawn is that they can function in both sorts of ways.[1] This conclusion is actually suggested by Newton's own presentation, where he calls them axioms *or* laws of motion; and it is easily accommodated on the present view. Here these expressions may be seen on the one hand to be definitions or axioms which, taken together, delineate the abstract *model*, in its simplest form, which underlies Newton's theory. Functioning in this way, Newton's axioms also tells us, for example, how we are to compute the values of certain functions (sometimes with the aid of explicit lemmas) in the context of the model, where bodies are conceived as mass-points having no extension. We note also that these particular expressions may be replaced by others, and from the present point of view the legitimacy of such a replacement in this context depends on whether the new expressions succeed in delineating a model in which the function values obtained by computation remain the same. On the other hand, the present view also suggests that these same expressions may be seen as *theoretical laws* (as discussed in Chapter 10), in which case they are not to be employed simply in the context of the model, but are rather to constitute the first step to be taken in applying the model to empirical reality.

With regard to this issue we see that on the set-theoretic conception Newton's three laws are treated as axioms, and all considerations are restricted to the nature of the relations that obtain between elements (particles) in the model. The Empiricist and Popperian views, on the

---

[1] Cf. e.g. Hanson (1958), pp. 97 ff., and Ellis (1965), pp. 61–62.

other hand, go in quite the other direction and see these laws as applying to empirical reality, ignoring altogether the model they serve to define. In this way then the present study may be seen as a sort of amalgam of these two views, for it recognizes the correctness of applying these laws or axioms in both directions.

2. KEPLER'S LAWS OF PLANETARY MOTION

One of the major achievements of Newton's theory of gravitation is considered to be its ability to explain Kepler's three laws of planetary motion. (Kepler published his first two laws in 1609, and his third in 1619.) Here we will investigate how, on the present view, Newton's theory may be seen as doing this, as well as why Kepler's laws taken together do not constitute a scientific theory. Kepler's laws may be presented as follows:[2]

1. The orbit of each planet is an ellipse, with the sun at one focus.
2. Each planet moves in its orbit at such a speed that a line joining it to the sun sweeps over equal areas in equal times.
3. The square of the time required for one revolution of a planet about the sun is proportional to the cube of its mean distance from the sun.

The reason usually given for Kepler's laws, taken together, not ranking as a theory is their unequivocally empirical character—i.e. the fact that they are 'instantial' in the sense that they refer to the individual planets in the solar system and, unlike a theory, are capable of being tested more or less directly.[3] The present view supports a distinction along these lines, and in fact provides an explanation of it, viz., that, unlike Newton's theory, Kepler's laws are not integrally related to a model, or, as Campbell might say, they do not evince any analogy, as they must in order to attain the status of theory.

The way in which Newton's theory explains each of Kepler's laws is in principle the same for each law, so we might simply restrict

---

[2] For the sake of readability none of the laws considered in the present chapter have been expressed as equations.

[3] For a distinction between laws and theories along this line, see Feyerabend (1962), p. 28 n. We note that, as remarked in earlier chapters, Popper's conception of science does not afford the making of this distinction, and that Popper himself treats Kepler's laws as constituting a scientific theory: cf. e.g. Popper (1957), p. 198.

ourselves to a consideration of how this transpires in the case of Kepler's first law. We note that on Newton's model, when only two masses (mass-points) are considered—one being appreciably greater than the other and taken as non-accelerating—if the smaller mass is in motion with respect to the larger it will describe a conic with the larger mass at a focus. And the nature of the conic—i.e. whether it be an ellipse or a parabola—will depend on the velocity with which the smaller mass is moving. In order to explain Kelper's first law, this situation obtaining in the abstract model must be applied, with the aid of correspondence rules, to the solar system. This is done by associating the sun with the non-accelerating mass, and an individual planet with the other mass; these conditions set, the planet's velocity is then independently determined by empirical means. Given the particular results of this determination, Newton's theory suggests that, other things being equal, the planet should describe a perfect ellipse about the sun, one focus of the ellipse being occupied by the sun itself. It is in this way then that Newton's theory serves to explain Kepler's first law.

In discussing the nature of the relation between Newton's gravitational theory and Kepler's laws, Duhem, and following him Popper, has suggested that Newton's theory actually contradicts Kepler's laws.[4] What they intend by this is that, as is readily accepted on the present view, when the masses of more than one planet are taken into account, as well as the gravitational force exerted by the planet on the sun, then Newton's theory will, with regard to Kepler's first law for example, not suggest that the planet in question should describe a perfect ellipse, but rather that it should describe some other curve, since its orbit will be perturbed by the effect of its own mass and those of the other planets. But their use of the term "contradiction" deserves comment, for if this term is used in its philosophically precise sense then one of the two contradicting assertions must be true, whereas in the Newton-Kepler case, when the masses of all the planets are taken into account it might well be that neither the situation suggested by Newton's theory nor that depicted by Kepler's laws actually obtains. And a second point in this regard is that if we

[4] Cf. Duhem (1906), p. 193, and Popper (1957), pp. 200ff.

flatly assert that the two systems contradict one another, not only is the fact obscured that Newton's theory and Kepler's laws are entities of essentially different sorts, but we will be hard put to account for the sense in which Newton's system may be seen to explain Kepler's. (This last point is pursued below in the context of Galileo's laws.)

The way in which Newton's theory serves to explain Kepler's first law has been outlined above. Now we will show in somewhat more detail how, on the present view, this comes about, and at the same time the way in which the two systems nevertheless conflict. The idea here is that the relation between Newton's theory and Kepler's law is essentially of the same sort as the relation between van der Waals' gas model and Boyle's law or the general gas law (note: not the ideal gas model), as presented in Chapter 10. And the key notion in this regard is that of reduction, as it was given in that earlier chapter. Thus we should say that, when the value of the mass parameter is taken to the limit zero in the case of the planets in the solar system other than the one with which we are presently concerned, and the gravitational force exerted by the planet on the sun is neglected, then Newton's theory suggests identical results to Kepler's first law, and in this way explains that law. But, as in the case of the volume of the molecules in a real gas, in applying Newton's theory in order to understand the motions of all the planets as actually observed, these parameters are given positive values. In this case then Newton's theory (with the aid of correspondence rules) and Kepler's first law respectively suggest different results from the employment of the same instruments (telescope etc.) in the plotting of the paths of the planets. And it is in this sense then that the two systems are incompatible.

3. GALILEO'S LAWS OF INERTIA AND FREE FALL

Galileo, in his *Dialogues Concerning Two New Sciences* (1638), gives two laws for the motion of bodies. The first to be presented here has two forms, one being that if the earth were a frictionless sphere, a body moving on it would continue in its motion forever. This law —Galileo's *law of inertia*—will here be treated in its second form however, which Galileo arrives at by a further process of idealization; in this form it may be presented as follows:

If a body is set in motion along a perfectly smooth horizontal plane, it will continue to move in such a way as is uniform and perpetual provided that the plane has no limits.

The second of Galileo's laws of motion is his *law for free fall*, which may be expressed as:

In falling freely toward the earth a body experiences constant acceleration with respect to the surface of the earth.

Here we note that in order to fall freely a body must be in a vacuum.

We might first consider whether Galileo's two laws, together, constitute a theory. On the present view it seems that they do, for these laws, in their explicit nullification of empirical factors, clearly depict an idealized model.[5] From this point of view, an infinite frictionless plane, for example, would exist in the model or theory, and that to which it corresponds in the physical world would be some local portion of the earth's surface. Thus we should say here that in this respect Galileo's laws of motion more closely resemble Newton's gravitational theory than do Kepler's laws.

On the basis of Galileo's two laws of motion we obtain what may be called a third law, which is that any projectile in motion relative to the surface of the earth (horizontal plane), when the effect of air resistance is neglected, will describe a parabola (the limiting case of which is a straight line). Newton suggests that this same result can be obtained on his theory if the centre of force (the centre of mass of the earth) is considered as being infinitely distant.[6] This of course is a counter-factual assumption similar to that regarding the absence of perturbing masses as is required in deriving Kepler's laws from Newton's theory. Thus we see that Newton's theory is capable of explaining Galileo's results concerning the motion of projectiles in the same sort of way as it is able to explain Kepler's laws, a particular instance of the distance parameter here being given a limiting (infinite) value; and the method employed is again one of reduction, as this term is understood in the present study.

---

[5] For comments on Galileo's two laws in this regard see Krajewski (1977), pp. 18 & 26. (We note however that it may be the case that the depiction of an idealized model need not in itself suffice for an expression's being that of a theory, though this question will not be pursued here.)

[6] Cf. the scholium following Proposition X, Problem V in Book I of Newton (1687).

In discussing the nature of the relation between Newton's theory and Kepler's and Galileo's laws respectively, Popper has said:

> Newton's theory unifies Galileo's and Kepler's. But far from being a mere conjunction of these two theories—which play the part of *explicanda* for Newton's—*it corrects them while explaining them*. ... Far from repeating its *explicandum*, the new theory contradicts it and corrects it.[7]

And later on the same page he makes reference to Bohr's 'principle of correspondence', which is essentially the same as the notion of reduction employed in the present study. But once again, as in Chapter 6, we have an instance of Popper presenting a relatively acceptable description of what occurs in actual science, but one which cannot be captured on his own conception. On Popper's view, as has been treated in detail in the earlier chapters of this study, explanation consists in deductive subsumption. Here, however, he wants to apply this notion not only to a case where such subsumption does not obtain, but to one which, on his own way of thinking, involves contradiction. Popper wants to have his cake and eat it too, and even goes so far as to provide the icing with his further suggestion that Newton's theory *corrects* its predecessors—a notion nowhere treated in the actual presentation of his conception of how competing theories are related to one another.

On the present conception however it may be said that, while Newton's theory, in application, is perspectivally incompatible with the 'theory' lying behind Galileo's laws, and suggests results differing both from those expected on the basis of Galileo's laws and from those suggested by Kepler's laws, it provides a unified account of their respective domains: planetary and terrestrial motion. It places both domains within one conceptual perspective. Following the Gestalt Model we may see that these unified domains can be presented in a way which is independent of the theories or laws concerned: just as the domain of application of the 'duck' and 'rabbit' concepts may be described as a line on a piece of paper, the realm of application of the above views may be said to consist in the data resulting from

[7] Popper (1957), p. 202.

measurements of position and velocity etc. of certain physical bodies. These data constitute a 'background of accepted fact' which, while not *absolute* in that its viability is based on certain empirical laws, is nevertheless *neutral* as regards the views under consideration. In this regard we might consider for example the data accumulated by Tycho Brahe on the basis of which Kepler discovered his three laws of planetary motion: these data are neutral to both Kepler's laws and Newton's theory, and constitute a realm to which each can be applied —in this way the data might conceivably have played an independent role in the debate between them. Similarly, neutral data would of course be available for an evaluation of Galileo's and Newton's theories.

Now, on this view it is not necessary that any of the above laws or theories should take full account of all of the relevant data. With the passage of time after the determination of initial conditions we might expect that Newton's and Galileo's theories and Kepler's laws would be completely off in their predictions (or retrodictions). Nevertheless, those predictions concerning terrestrial and planetary motion made on the basis of Newton's theory give numerical values *closer* to the values obtained using instruments in actual measurements. On the basis of such evidence, in conjunction with the fact that Newton's theory provides a unified or simple account having a broader scope than either of its rivals, we may say that scientific progress has been made in moving from the laws of Kepler and Galileo to the theory of Newton.

But the superiority of Newton's theory in this respect should not mean the utter dismissal of its rivals. Kepler's laws, for example, afford a description of planetary motion which is sufficiently accurate for many purposes, and which has the advantage of being conceptually more manageable that the account provided by Newton. Nor does this superiority suggest that Galileo's and Kepler's laws are false, while Newton's theory is true, or more true. Rather, it implies that, of the three, Newton's theory is the best *applicable* to the domain of terrestrial and planetary motion.

Thus at the same time as affording a conception of theory conflict as perspectival incompatibility, and incommensurability as shift of conceptual perspective, the present conception avoids relativism through the provision of a realistic notion of scientific progress as depending on the results of measurement. It may be kept in mind though, that while the above might function as a criterion as to what should constitute progress in science, it need be no guide as to what in any broader realm should constitute epistemological advance.

REFERENCES

Achinstein, P.

(1964) "On the Meaning of Scientific Terms", *Journal of Philosophy* **61**, 1964.

(1965) "The Problem of Theoretical Terms", *American Philosophical Quarterly* **2**, 1965.

Adams, E. W.

(1959) "The Foundations of Rigid Body Mechanics and the Derivation of Its Laws from those of Particle Mechanics", *The Axiomatic Method*, L. Henkin et al. (eds.), North-Holland, Amsterdam, 1959.

Agazzi, E.

(1976) "The Concept of Empirical Data", *Formal Methods in the Methodology of Empirical Sciences,* M. Przełecki et al. (eds.), D. Reidel, Dordrecht, 1976.

(1977*a*) "Subjectivity, Objectivity and Onotological Commitment in the Empirical Sciences", *Historical and Philosophical Dimensions of Logic, Methodology and Philosophy of Science*, Butts and Hintikka (eds.), D. Reidel, Dordrecht, 1977.

(1977*b*) "The Role of Metaphysics in Contemporary Philosophy", *Ratio* **XIX**, 1977.

(1978) "L'Objectivité Scientifique: Est-Elle Possible Sans La Mesure?", *Diogène* **104**, 1978.

Allen, H. S. and Maxwell, R. S.

(1939) *A Text-Book of Heat*, Part I, Macmillan and Co., London, 1944.

Andersson, J. and Furberg, M.

(1966) *Språk och Påverkan*, Bokförlaget Aldus/Bonniers, Stockholm, 1974.

Aristotle

The Works of Aristotle, Vol. 1, W. D. Ross (ed.), Oxford University Press, London, 1928.

Ayer, A. J.

(1936) *Language, Truth and Logic*, Dover Publications, New York, 1952.

Barton, A. W.

(1933) *A Text Book on Heat*, Longmans, Green and Co., London, 1944.

Bocheński, I. M.

(1961) *A History of Formal Logic*, University of Notre Dame Press, Notre Dame, Indiana, 1961.

Boltzmann, L.

(1896) *Lectures on Gas Theory*, S. G. Brush (tr.), University of California Press, 1964.

Bridgman, P. W.
(1936) *The Nature of Physical Theory*, Princeton University Press, Princeton, 1936.
Campbell, N. R.
(1920) *Physics: The Elements*, Cambridge University Press, Cambridge, 1920; reprinted as *Foundations of Science*, Dover Publications, New York, 1957.
(1921) *What is Science?*, Dover Publications, New York, 1953.
(1928) *An Account of the Principles of Measurement and Calculation*, Longmans, Green and Co., London, 1928.
Carnap, R.
(1936) *Testability and Meaning*, Whitlock's, New Haven, 1954.
(1966*a*) *Philosophical Foundations of Physics*, Basic Books, New York, 1966.
(1966*b*) "Probability and Content Measure", *Mind, Matter, and Method*, P. Feyerabend and G. Maxwell (eds.), University of Minnesota Press, Minneapolis, 1966.
Cohen, M. and Nagel, E.
(1934) *An Introduction to Logic and Scientific Method*, Harcourt Brace and World, New York, 1934.
Deutscher, M.
(1968) "Popper's Problem of an Empirical Basis", *Australasian Journal of Philosophy* **46**, 1968.
Dilworth, C.
(1978) "On the Nature of the Relation Between Successive Scientific Theories", *Epistemologia* **1**, 1978.
(1979) "Correspondence Principle and Growth of Science" (Review of Krajewski (1977)), *Epistemologia* **2**, 1979.
Dilworth, C./Bunge, M.
(1979) "On Bunge's *Treatise on Basic Philosophy*", *Epistemologia* **2**, 1979.
Duhem, P.
(1906) *The Aim and Structure of Physical Theory*, P. P. Wiener (tr.), Atheneum, New York, 1962.
Ellis, B.
(1965) "The Origin and Nature of Newton's Laws of Motion", *Beyond the Edge of Certainty*, R. G. Colodny (ed.), Prentice Hall, N.J., 1965.
Feyerabend, P. K.
(1962) "Explanation, Reduction, and Empiricism", *Minnesota Studies in the Philosophy of Science* **III**, University of Minnesota Press, Minneapolis, 1962.
(1963) "How to be a Good Empiricist—A Plea for Tolerance in Matters Epistemological", *The Philosophy of Science*, P. H. Nidditch (ed.), Oxford University Press, London, 1968.

(1965 *a*) "On the "Meaning" of Scientific Terms", *Journal of Philosophy* **62**, 1965.
(1965 *b*) "Reply to Criticism", *In Honor of Philipp Frank*, R. S. Cohen and M. W. Wartofsky (eds.), D. Reidel, Dordrecht, 1965.
(1970) "Consolations for the Specialist", *Criticism and the Growth of Knowledge*, I. Lakatos and A. Musgrave (eds.), Cambridge University Press, Cambridge, 1970.
(1974) "Popper's *Objective Knowledge*", *Inquiry* **17**, 1974.
(1975) *Against Method*, New Left Books, London, 1975.
(1977) "Changing Patterns of Reconstruction", *British Journal for the Philosophy of Science* **28**, 1977.
(1978) *Science in a Free Society*, New Left Books, London, 1978.

Fürth, R.
(1969) "The Role of Models in Theoretical Physics", *Proceedings of the Boston Colloquium for the Philosophy of Science 1966–1968*, R. S. Cohen and M. W. Wartofsky (eds.), D. Reidel, Dordrecht, 1969.

Galilei, G.
(1638) *Dialogues Concerning Two New Sciences*, H. Crew and A. de Salvio (trs.), Dover Publications, New York, 1954.

Giedymin, J.
(1970) "The Paradox of Meaning Variance", *British Journal for the Philosophy of Science* **21**, 1970.

Haack, S.
(1976) "'Is it True What They Say About Tarski?'", *Philosophy* **51**, 1976.

Hanson, N. R.
(1958) *Patterns of Discovery*, Cambridge University Press, Cambridge, 1958.
(1966) "Equivalence: The Paradox of Theoretical Analysis", *Mind, Matter, and Method*, P. Feyerabend and G. Maxwell (eds.), University of Minnesota Press, Minneapolis, 1966.

Hanson, N. R. et al.
(1970) "Discussion at the Conference on Correspondence Rules", *Minnesota Studies in the Philosophy of Science* **IV**, M. Radner and S. Winokur (eds.), University of Minnesota Press, Minneapolis, 1970.

Hempel, C. G.
(1962) "Deductive-Nomological vs. Statistical Explanation", *Minnesota Studies in the Philosophy of Science* **III**, H. Feigl and G. Maxwell (eds.), University of Minnesota Press, Minneapolis, 1962.
(1965) *Aspects of Scientific Explanation*, Free Press, New York, 1965.
(1966) "Recent Problems of Induction", *Mind and Cosmos*, R. G. Colodny (ed.), University of Pittsburg Press, Pittsburg, 1966.
(1970) "On the "Standard Conception" of Scientific Theories", *Minnesota Studies in the Philosophy of Science* **IV**, M. Radner and S. Winokur (eds.), University of Minnesota Press, Minneapolis, 1970.

Hempel, C. G. and Oppenheim, P.
(1948) "Studies in the Logic of Explanation", *Philosophy of Science* **15**, 1948.
Hesse, M.
(1963) "A New Look at Scientific Explanation", *Review of Metaphysics* **17**, 1963–1964.
(1966) *Models and Analogies in Science*, University of Notre Dame Press, Notre Dame, Indiana, 1966.
Johansson, I.
(1978) "Hönan och Galvanometern", *Några Aktuella Vetenskapsfilosofiska Problem*, Umeå Studies in the Theory and Philosophy of Science, No. 14, 1978.
Jordan, P.
(1944) *Physics of the 20th Century*, E. Oshry (tr.), Philosophical Library, New York, 1944.
Kaiser, M.
(1979) "On Sneed and Rationality", paper presented at Inter-University Centre, Dubrovnik, April, 1979.
Kenner, L.
(1965) "The Triviality of the Red-Green Problem", *Analysis* **25**, 1964–1965.
Keynes, J. M.
(1921) *A Treatise on Probability*, Macmillan and Co., London, 1921.
Kneale, W.
(1964) "On Popper's Use of the Notion of Absolute Logical Probability", *The Critical Approach to Science and Philosophy*, M. Bunge (ed.), Free Press of Glencoe, London, 1964.
Krajewski, W.
(1977) *Correspondence Principle and Growth of Science*, D. Reidel, Dordrecht, 1977.
Kuhn, T. S.
(1957) *The Copernican Revolution*, Harvard University Press, Cambridge, Mass., 1957.
(1961) "The Function of Measurement in Modern Physical Science", *Isis* **52**, 1961.
(1962) *The Structure of Scientific Revolutions*, 1970; Vol. 2, No. 2, International Encyclopedia of Unified Science, University of Chicago Press, Chicago.
(1970a) "Postscript—1969" in Kuhn (1962).
(1970b) "Logic of Discovery or Psychology of Research?", *Criticism and the Growth of Knowledge*, Cambridge University Press, Cambridge, 1970.
(1970c) "Reflections on My Critics", *Criticism and the Growth of Knowledge*, Cambridge University Press, Cambridge, 1970.
(1971) "Notes on Lakatos", *PSA 1970: In Memory of Rudolf Carnap*, R. C. Buck and R. S. Cohen (eds.), D. Reidel, Dordrecht, 1971.

9†–812581 Dilworth

(1974) "Second Thoughts on Paradigms", *The Structure of Scientific Theories*, F. Suppe (ed.), University of Illinois Press, Urbana, 1974.
(1976) "Theory-Change as Structure-Change: Comments on the Sneed Formalism", *Erkenntnis* **10**, 1976.
(1977) *The Essential Tension*, University of Chicago Press, Chicago, 1977.

Lakatos, I.
(1968) "Changes in the Problem of Inductive Logic", *The Problem of Inductive Logic*, I. Lakatos (ed.), North-Holland, Amsterdam, 1968.
(1970) "Falsification and the Methodology of Scientific Research Programmes", *Criticism and the Growth of Knowledge*, I. Lakatos and A. Musgrave (eds.), Cambridge University Press, Cambridge, 1970.

Laudan, L.
(1977) *Progress and Its Problems*, University of California Press, Berkeley, 1977.

Martin, M.
(1971) "Referential Variance and Scientific Objectivity", *British Journal for the Philosophy of Science* **22**, 1971.

Mayr, D.
(1976) "Investigations of the Concept of Reduction I", *Erkenntnis* **10**, 1976.

McKinsey, J. C. C., Sugar, A. C., and Suppes, P.
(1953) "Axiomatic Foundations of Classical Particle Mechanics", *Journal of Rational Mechanics and Analysis* **2**, 1953.

Miller, D. W.
(1974) "Popper's Qualitative Theory of Verisimilitude", *British Journal for the Philosophy of Science* **25**, 1974.

Mitton, R. G.
(1939) *Heat*, J. M. Dent and Sons, London, 1945.

Moulines, C. U.
(1976) "Approximate Application of Empirical Theories: A General Explication", *Erkenntnis* **10**, 1976.

Nagel, E.
(1961) *The Structure of Science*, Harcourt, Brace & World, New York, 1961.

Newton, I.
(1687) *Mathematical Principles of Natural Philosophy*, A. Motte (tr.), University of California Press, Berkeley, 1947.

Nowak, L.
(1979) "Idealization and Rationalization", *Epistemologia* **2**, Special Issue, 1979.
(1980) *The Structure of Idealization*, D. Reidel, Dordrecht, 1980.

O'Connor, D. J.
(1955) "Incompatible Properties", *Analysis* **15**, 1955.

Partington, J. R.

(1961) *A History of Chemistry*, Vol. 2, Macmillan and Co., London, 1961.

Poincaré, H.

(1903) *Science and Hypothesis*, Dover Publications, New York, 1952.

Popper, K. R.

(1934) *The Logic of Scientific Discovery*, Hutchinson & Co., London, 1959.

(1949) "The Bucket and the Searchlight: Two Theories of Knowledge" in Popper (1973).

(1957*a*) "The Aim of Science" in Popper (1973).

(1957*b*) "Science: Conjectures and Refutations" in Popper (1962).

(1958) "On the Status of Science and Metaphysics" in Popper (1962).

(1959) "New Appendicies" (and new footnotes) in Popper (1934).

(1962) *Conjectures and Refutations*, Harper & Row, New York, 1963.

(1972) *Objective Knowledge*, Oxford University Press, Oxford, 1972.

(1973) *Objective Knowledge* (printed with corrections), Oxford University Press, Oxford, 1973.

(1975) "The Rationality of Scientific Revolutions", *Problems of Scientific Revolution*, R. Harré (ed.), Clarendon Press, Oxford, 1975.

Putnam, H.

(1962) "What Theories Are Not", *Theories and Observation in Science*, R. E. Grandy (ed.), Prentice-Hall, Engelwood Cliffs, N.J., 1973.

Ramsey, F. P.

(1931) "Theories" in his *The Foundations of Mathematics*, Kegan Paul, Trench, Trubner & Co., London, 1931.

Russell, B.

(1940) *An Inquiry into Meaning and Truth*, W. W. Norton & Co., New York, 1940.

Scriven, M.

(1962) "Explanations, Predictions, and Laws", *Minnesota Studies in the Philosophy of Science* **III**, H. Feigl and G. Maxwell (eds.), Univeristy of Minnesota Press, Minneapolis, 1962.

Shapere, D.

(1966) "Meaning and Scientific Change", *Mind and Cosmos*, R. G. Colodny (ed.), University of Pittsburg Press, Pittsburg, 1966.

Sneed, J. D.

(1971) *The Logical Structure of Mathematical Physics*, D. Reidel, Dordrecht: Holland, 1979.

(1976) "Philosophical Problems in the Empirical Science of Science: A Formal Approach", *Erkenntnis* **10**, 1976.

Spector, M.

(1965) "Models and Theories", *British Journal for the Philosophy of Science* **16**, 1965–1966.

Stegmüller, W.
(1973) *The Structure and Dynamics of Theories*, Springer-Verlag, New York, 1976.
(1976) ''Accidental ('Non-Substantial') Theory Change and Theory Dislodgment: To What Extent Logic Can Contribute to a Better Understanding of Certain Phenomena in the Dynamics of Theories'', *Erkenntnis* **10**, 1976.
(1979) *The Structuralist View of Theories,* Springer-Verlag, Berlin, 1979.
Suppes, P.
(1967) *Set-Theoretical Structures in Science*, stencil, Stanford University, Stanford, 1970.
Tichý, P.
(1974) ''On Popper's Definitions of Verisimilitude'', *British Journal for the Philosophy of Science* **25**, 1974.
Ullman, S.
(1962) *Semantics*, Basil Blackwell, Oxford, 1967.
Wallace, W. A.
(1974) *Causality and Scientific Explanation*, Vol. 2, University of Michigan Press, Ann Arbor, 1974.
Wartofsky, M. W.
(1968) *Conceptual Foundations of Scientific Thought*, Macmillan Co., New York, 1968.
Wittgenstein, L.
(1921) *Tractatus Logico-Philosophicus*, D. F. Pears and B. F. McGuinness (trs.), Routledge & Kegan Paul, London, 1961.
(1953) *Philosophical Investigations*, G. E. M. Anscombe (tr.), Basil Blackwell, Oxford, 1972.

INDEX

SYNTHESE LIBRARY

SYNTHESE LIBRARY

Studies in Epistemology, Logic, Methodology,
and Philosophy of Science

1. J. M. Bocheński, *A Precis of Mathematical Logic.* 1959.
2. P. L. Guiraud, *Problèmes et méthodes de la statistique linguistique.* 1960.
3. Hans Freudenthal (ed.), *The Concept and the Role of the Model in Mathematics and Natural and Social Sciences.* 1961.
4. Evert W. Beth, *Formal Methods. An Introduction to Symbolic Logic and the Study of Effective Operations in Arithmetic and Logic.* 1962.
5. B. H. Kazemier and D. Vuysje (eds.), *Logic and Language. Studies Dedicated to Professor Rudolf Carnap on the Occasion of His Seventieth Birthday.* 1962.
6. Marx W. Wartofsky (ed.), *Proceedings of the Boston Colloquium for the Philosophy of Science 1961-1962.* Boston Studies in the Philosophy of Science, Volume I. 1963.
7. A. A. Zinov'ev, *Philosophical Problems of Many-Valued Logic.* 1963.
8. Georges Gurvitch, *The Spectrum of Social Time.* 1964.
9. Paul Lorenzen, *Formal Logic.* 1965.
10. Robert S. Cohen and Marx W. Wartofsky (eds.), *In Honor of Philipp Frank.* Boston Studies in the Philosophy of Science, Volume II. 1965.
11. Evert W. Beth, *Mathematical Thought. An Introduction to the Philosophy of Mathematics.* 1965.
12. Evert W. Beth and Jean Piaget, *Mathematical Epistemology and Psychology.* 1966.
13. Guido Küng, *Ontology and the Logistic Analysis of Language. An Enquiry into the Contemporary Views on Universals.* 1967.
14. Robert S. Cohen and Marx W. Wartofsky (eds.), *Proceedings of the Boston Colloquium for the Philosophy of Science 1964-1966. In Memory of Norwood Russell Hanson.* Boston Studies in the Philosophy of Science, Volume III. 1967.
15. C. D. Broad, *Induction, Probability, and Causation. Selected Papers.* 1968.
16. Günther Patzig, *Aristotle's Theory of the Syllogism. A Logical-Philosophical Study of Book A of the Prior Analytics.* 1968.
17. Nicholas Rescher, *Topics in Philosophical Logic.* 1968.
18. Robert S. Cohen and Marx W. Wartofsky (eds.), *Proceedings of the Boston Colloquium for the Philosophy of Science 1966-1968.* Boston Studies in the Philosophy of Science, Volume IV. 1969.

19. Robert S. Cohen and Marx W. Wartofsky (eds.), *Proceedings of the Boston Colloquium for the Philosophy of Science 1966-1968.* Boston Studies in the Philosophy of Science, Volume V. 1969.
20. J. W. Davis, D. J. Hockney, and W. K. Wilson (eds.), *Philosophical Logic.* 1969.
21. D. Davidson and J. Hintikka (eds.), *Words and Objections. Essays on the Work of W. V. Quine.* 1969.
22. Patrick Suppes, *Studies in the Methodology and Foundations of Science. Selected Papers from 1911 to 1969.* 1969.
23. Jaakko Hintikka, *Models for Modalities. Selected Essays.* 1969.
24. Nicholas Rescher *et al.* (eds.), *Essays in Honor of Carl G. Hempel. A Tribute on the Occasion of His Sixty-Fifth Birthday.* 1969.
25. P. V. Tavanec (ed.), *Problems of the Logic of Scientific Knowledge.* 1969.
26. Marshall Swain (ed.), *Induction, Acceptance, and Rational Belief.* 1970.
27. Robert S. Cohen and Raymond J. Seeger (eds.), *Ernst Mach: Physicist and Philosopher.* Boston Studies in the Philosophy of Science, Volume VI. 1970.
28. Jaakko Hintikka and Patrick Suppes, *Information and Inference.* 1970.
29. Karel Lambert, *Philosophical Problems in Logic. Some Recent Developments.* 1970.
30. Rolf A. Eberle, *Nominalistic Systems.* 1970.
31. Paul Weingartner and Gerhard Zecha (eds.), *Induction, Physics, and Ethics.* 1970.
32. Evert W. Beth, *Aspects of Modern Logic.* 1970.
33. Risto Hilpinen (ed.), *Deontic Logic: Introductory and Systematic Readings.* 1971.
34. Jean-Louis Krivine, *Introduction to Axiomatic Set Theory.* 1971.
35. Joseph D. Sneed, *The Logical Structure of Mathematical Physics.* 1971.
36. Carl R. Kordig, *The Justification of Scientific Change.* 1971.
37. Milic Capek, *Bergson and Modern Physics.* Boston Studies in the Philosophy of Science, Volume VII. 1971.
38. Norwood Russell Hanson, *What I Do Not Believe, and Other Essays* (ed. by Stephen Toulmin and Harry Woolf). 1971.
39. Roger C. Buck and Robert S. Cohen (eds.), *PSA 1970. In Memory of Rudolf Carnap.* Boston Studies in the Philosophy of Science, Volume VIII. 1971.
40. Donald Davidson and Gilbert Harman (eds.), *Semantics of Natural Language.* 1972.
41. Yehoshua Bar-Hillel (ed.), *Pragmatics of Natural Languages.* 1971.
42. Sören Stenlund, *Combinators, λ-Terms and Proof Theory.* 1972.
43. Martin Strauss, *Modern Physics and Its Philosophy. Selected Papers in the Logic, History, and Philosophy of Science.* 1972.
44. Mario Bunge, *Method, Model and Matter.* 1973.
45. Mario Bunge, *Philosophy of Physics.* 1973.
46. A. A. Zinov'ev, *Foundations of the Logical Theory of Scientific Knowledge (Complex Logic).* (Revised and enlarged English edition with an appendix by G. A. Smirnov, E. A. Sidorenka, A. M. Fedina, and L. A. Bobrova.) Boston Studies in the Philosophy of Science, Volume IX. 1973.
47. Ladislav Tondl, *Scientific Procedures.* Boston Studies in the Philosophy of Science, Volume X. 1973.
48. Norwood Russell Hanson, *Constellations and Conjectures* (ed. by Willard C. Humphreys, Jr.). 1973.

49. K. J. J. Hintikka, J. M. E. Moravcsik, and P. Suppes (eds.), *Approaches to Natural Language.* 1973.
50. Mario Bunge (ed.), *Exact Philosophy – Problems, Tools, and Goals.* 1973.
51. Radu J. Bogdan and Ilkka Niiniluoto (eds.), *Logic, Language, and Probability.* 1973.
52. Glenn Pearce and Patrick Maynard (eds.), *Conceptual Change.* 1973.
53. Ilkka Niiniluoto and Raimo Tuomela, *Theoretical Concepts and Hypothetico-Inductive Inference.* 1973.
54. Roland Fraissé, *Course of Mathematical Logic* – Volume 1: *Relation and Logical Formula.* 1973.
55. Adolf Grünbaum, *Philosophical Problems of Space and Time.* (Second, enlarged edition.) Boston Studies in the Philosophy of Science, Volume XII. 1973.
56. Patrick Suppes (ed.), *Space, Time, and Geometry.* 1973.
57. Hans Kelsen, *Essays in Legal and Moral Philosophy* (selected and introduced by Ota Weinberger). 1973.
58. R. J. Seeger and Robert S. Cohen (eds.), *Philosophical Foundations of Science.* Boston Studies in the Philosophy of Science, Volume XI. 1974.
59. Robert S. Cohen and Marx W. Wartofsky (eds.), *Logical and Epistemological Studies in Contemporary Physics.* Boston Studies in the Philosophy of Science, Volume XIII. 1973.
60. Robert S. Cohen and Marx W. Wartofsky (eds.), *Methodological and Historical Essays in the Natural and Social Sciences. Proceedings of the Boston Colloquium for the Philosophy of Science 1969-1972.* Boston Studies in the Philosophy of Science, Volume XIV. 1974.
61. Robert S. Cohen, J. J. Stachel, and Marx W. Wartofsky (eds.), *For Dirk Struik. Scientific, Historical and Political Essays in Honor of Dirk J. Struik.* Boston Studies in the Philosophy of Science, Volume XV. 1974.
62. Kazimierz Ajdukiewicz, *Pragmatic Logic* (transl. from the Polish by Olgierd Wojtasiewicz). 1974.
63. Sören Stenlund (ed.), *Logical Theory and Semantic Analysis. Essays Dedicated to Stig Kanger on His Fiftieth Birthday.* 1974.
64. Kenneth F. Schaffner and Robert S. Cohen (eds.), *Proceedings of the 1972 Biennial Meeting, Philosophy of Science Association.* Boston Studies in the Philosophy of Science, Volume XX. 1974.
65. Henry E. Kyburg, Jr., *The Logical Foundations of Statistical Inference.* 1974.
66. Marjorie Grene, *The Understanding of Nature. Essays in the Philosophy of Biology.* Boston Studies in the Philosophy of Science, Volume XXIII. 1974.
67. Jan M. Broekman, *Structuralism: Moscow, Prague, Paris.* 1974.
68. Norman Geschwind, *Selected Papers on Language and the Brain.* Boston Studies in the Philosophy of Science, Volume XVI. 1974.
69. Roland Fraissé, *Course of Mathematical Logic* – Volume 2: *Model Theory.* 1974.
70. Andrzej Grzegorczyk, *An Outline of Mathematical Logic. Fundamental Results and Notions Explained with All Details.* 1974.
71. Franz von Kutschera, *Philosophy of Language.* 1975.
72. Juha Manninen and Raimo Tuomela (eds.), *Essays on Explanation and Understanding. Studies in the Foundations of Humanities and Social Sciences.* 1976.

73. Jaakko Hintikka (ed.), *Rudolf Carnap, Logical Empiricist. Materials and Perspectives.* 1975.
74. Milic Capek (ed.), *The Concepts of Space and Time. Their Structure and Their Development.* Boston Studies in the Philosophy of Science, Volume XXII. 1976.
75. Jaakko Hintikka and Unto Remes, *The Method of Analysis. Its Geometrical Origin and Its General Significance.* Boston Studies in the Philosophy of Science, Volume XXV. 1974.
76. John Emery Murdoch and Edith Dudley Sylla, *The Cultural Context of Medieval Learning.* Boston Studies in the Philosophy of Science, Volume XXVI. 1975.
77. Stefan Amsterdamski, *Between Experience and Metaphysics. Philosophical Problems of the Evolution of Science.* Boston Studies in the Philosophy of Science, Volume XXXV. 1975.
78. Patrick Suppes (ed.), *Logic and Probability in Quantum Mechanics.* 1976.
79. Hermann von Helmholtz: *Epistemological Writings. The Paul Hertz/Moritz Schlick Centenary Edition of 1921 with Notes and Commentary by the Editors.* (Newly translated by Malcolm F. Lowe. Edited, with an Introduction and Bibliography, by Robert S. Cohen and Yehuda Elkana.) Boston Studies in the Philosophy of Science, Volume XXXVII. 1977.
80. Joseph Agassi, *Science in Flux.* Boston Studies in the Philosophy of Science, Volume XXVIII. 1975.
81. Sandra G. Harding (ed.), *Can Theories Be Refuted? Essays on the Duhem-Quine Thesis.* 1976.
82. Stefan Nowak, *Methodology of Sociological Research. General Problems.* 1977.
83. Jean Piaget, Jean-Blaise Grize, Alina Szeminska, and Vinh Bang, *Epistemology and Psychology of Functions.* 1977.
84. Marjorie Grene and Everett Mendelsohn (eds.), *Topics in the Philosophy of Biology.* Boston Studies in the Philosophy of Science, Volume XXVII. 1976.
85. E. Fischbein, *The Intuitive Sources of Probabilistic Thinking in Children.* 1975.
86. Ernest W. Adams, *The Logic of Conditionals. An Application of Probability to Deductive Logic.* 1975.
87. Marian Przelecki and Ryszard Wójcicki (eds.), *Twenty-Five Years of Logical Methodology in Poland.* 1977.
88. J. Topolski, *The Methodology of History.* 1976.
89. A. Kasher (ed.), *Language in Focus: Foundations, Methods and Systems. Essays Dedicated to Yehoshua Bar-Hillel.* Boston Studies in the Philosophy of Science, Volume XLIII. 1976.
90. Jaakko Hintikka, *The Intentions of Intentionality and Other New Models for Modalities.* 1975.
91. Wolfgang Stegmüller, *Collected Papers on Epistemology, Philosophy of Science and History of Philosophy.* 2 Volumes. 1977.
92. Dov M. Gabbay, *Investigations in Modal and Tense Logics with Applications to Problems in Philosophy and Linguistics.* 1976.
93. Radu J. Bogdan, *Local Induction.* 1976.
94. Stefan Nowak, *Understanding and Prediction. Essays in the Methodology of Social and Behavioral Theories.* 1976.
95. Peter Mittelstaedt, *Philosophical Problems of Modern Physics.* Boston Studies in the Philosophy of Science, Volume XVIII. 1976.

96. Gerald Holton and William Blanpied (eds.), *Science and Its Public: The Changing Relationship.* Boston Studies in the Philosophy of Science, Volume XXXIII. 1976.
97. Myles Brand and Douglas Walton (eds.), *Action Theory.* 1976.
98. Paul Gochet, *Outline of a Nominalist Theory of Proposition. An Essay in the Theory of Meaning.* 1980.
99. R. S. Cohen, P. K. Feyerabend, and M. W. Wartofsky (eds.), *Essays in Memory of Imre Lakatos.* Boston Studies in the Philosophy of Science, Volume XXXIX. 1976.
100. R. S. Cohen and J. J. Stachel (eds.), *Selected Papers of Léon Rosenfeld.* Boston Studies in the Philosophy of Science, Volume XXI. 1978.
101. R. S. Cohen, C. A. Hooker, A. C. Michalos, and J. W. van Evra (eds.), *PSA 1974: Proceedings of the 1974 Biennial Meeting of the Philosophy of Science Association.* Boston Studies in the Philosophy of Science, Volume XXXII. 1976.
102. Yehuda Fried and Joseph Agassi, *Paranoia: A Study in Diagnosis.* Boston Studies in the Philosophy of Science, Volume L. 1976.
103. Marian Przelecki, Klemens Szaniawski, and Ryszard Wójcicki (eds.), *Formal Methods in the Methodology of Empirical Sciences.* 1976.
104. John M. Vickers, *Belief and Probability.* 1976.
105. Kurt H. Wolff, *Surrender and Catch: Experience and Inquiry Today.* Boston Studies in the Philosophy of Science, Volume LI. 1976.
106. Karel Kosík, *Dialectics of the Concrete.* Boston Studies in the Philosophy of Science, Volume LII. 1976.
107. Nelson Goodman, *The Structure of Appearance.* (Third edition.) Boston Studies in the Philosophy of Science, Volume LIII. 1977.
108. Jerzy Giedymin (ed.), *Kazimierz Ajdukiewicz: The Scientific World-Perspective and Other Essays, 1931-1963.* 1978.
109. Robert L. Causey, *Unity of Science.* 1977.
110. Richard E. Grandy, *Advanced Logic for Applications.* 1977.
111. Robert P. McArthur, *Tense Logic.* 1976.
112. Lars Lindahl, *Position and Change. A Study in Law and Logic.* 1977.
113. Raimo Tuomela, *Dispositions.* 1978.
114. Herbert A. Simon, *Models of Discovery and Other Topics in the Methods of Science.* Boston Studies in the Philosophy of Science, Volume LIV. 1977.
115. Roger D. Rosenkrantz, *Inference, Method and Decision.* 1977.
116. Raimo Tuomela, *Human Action and Its Explanation. A Study on the Philosophical Foundations of Psychology.* 1977.
117. Morris Lazerowitz, *The Language of Philosophy. Freud and Wittgenstein.* Boston Studies in the Philosophy of Science, Volume LV. 1977.
118. Stanislaw Leśniewski, *Collected Works* (ed. by S. J. Surma, J. T. J. Srzednicki, and D. I. Barnett, with an annotated bibliography by V. Frederick Rickey). 1982. (Forthcoming.)
119. Jerzy Pelc, *Semiotics in Poland, 1894-1969.* 1978.
120. Ingmar Pörn, *Action Theory and Social Science. Some Formal Models.* 1977.
121. Joseph Margolis, *Persons and Minds. The Prospects of Nonreductive Materialism.* Boston Studies in the Philosophy of Science, Volume LVII. 1977.
122. Jaakko Hintikka, Ilkka Niiniluoto, and Esa Saarinen (eds.), *Essays on Mathematical and Philosophical Logic.* 1978.
123. Theo A. F. Kuipers, *Studies in Inductive Probability and Rational Expectation.* 1978.

124. Esa Saarinen, Risto Hilpinen, Ilkka Niiniluoto, and Merrill Provence Hintikka (eds.), *Essays in Honour of Jaakko Hintikka on the Occasion of His Fiftieth Birthday.* 1978.
125 Gerard Radnitzky and Gunnar Andersson (eds.), *Progress and Rationality in Science.* Boston Studies in the Philosophy of Science, Volume LVIII. 1978.
126. Peter Mittelstaedt, *Quantum Logic.* 1978.
127. Kenneth A. Bowen, *Model Theory for Modal Logic. Kripke Models for Modal Predicate Calculi.* 1978.
128. Howard Alexander Bursen, *Dismantling the Memory Machine. A Philosophical Investigation of Machine Theories of Memory.* 1978.
129. Marx W. Wartofsky, *Models: Representation and the Scientific Understanding.* Boston Studies in the Philosophy of Science, Volume XLVIII. 1979.
130. Don Ihde, *Technics and Praxis. A Philosophy of Technology.* Boston Studies in the Philosophy of Science, Volume XXIV. 1978.
131. Jerzy J. Wiatr (ed.), *Polish Essays in the Methodology of the Social Sciences.* Boston Studies in the Philosophy of Science, Volume XXIX. 1979.
132. Wesley C. Salmon (ed.), *Hans Reichenbach: Logical Empiricist.* 1979.
133. Peter Bieri, Rolf-P. Horstmann, and Lorenz Krüger (eds.), *Transcendental Arguments in Science. Essays in Epistemology.* 1979.
134. Mihailo Marković and Gajo Petrović (eds.), *Praxis. Yugoslav Essays in the Philosophy and Methodology of the Social Sciences.* Boston Studies in the Philosophy of Science, Volume XXXVI. 1979.
135. Ryszard Wójcicki, *Topics in the Formal Methodology of Empirical Sciences.* 1979.
136. Gerard Radnitzky and Gunnar Andersson (eds.), *The Structure and Development of Science.* Boston Studies in the Philosophy of Science, Volume LIX. 1979.
137. Judson Chambers Webb. *Mechanism, Mentalism, and Metamathematics. An Essay on Finitism.* 1980.
138. D. F. Gustafson and B. L. Tapscott (eds.), *Body, Mind, and Method. Essays in Honor of Virgil C. Aldrich.* 1979.
139. Leszek Nowak, *The Structure of Idealization. Towards a Systematic Interpretation of the Marxian Idea of Science.* 1979.
140. Chaim Perelman, *The New Rhetoric and the Humanities. Essays on Rhetoric and Its Applications.* 1979.
141. Wlodzimierz Rabinowicz, *Universalizability. A Study in Morals and Metaphysics.* 1979.
142. Chaim Perelman, *Justice, Law, and Argument. Essays on Moral and Legal Reasoning.* 1980.
143. Stig Kanger and Sven Öhman (eds.), *Philosophy and Grammar. Papers on the Occasion of the Quincentennial of Uppsala University.* 1981.
144. Tadeusz Pawlowski, *Concept Formation in the Humanities and the Social Sciences.* 1980.
145. Jaakko Hintikka, David Gruender, and Evandro Agazzi (eds.), *Theory Change, Ancient Axiomatics, and Galileo's Methodology. Proceedings of the 1978 Pisa Conference on the History and Philosophy of Science, Volume I.* 1981.
146. Jaakko Hintikka, David Gruender, and Evandro Agazzi (eds.), *Probabilistic Thinking, Thermodynamics, and the Interaction of the History and Philosophy of*

Science. Proceedings of the 1978 Pisa Conference on the History and Philosophy of Science, Volume II. 1981.

147. Uwe Mönnich (ed.), *Aspects of Philosophical Logic. Some Logical Forays into Central Notions of Linguistics and Philosophy.* 1981.
148. Dov M. Gabbay, *Semantical Investigations in Heyting's Intuitionistic Logic.* 1981.
149. Evandro Agazzi (ed.), *Modern Logic – A Survey. Historical, Philosophical, and Mathematical Aspects of Modern Logic and its Applications.* 1981.
150. A. F. Parker-Rhodes, *The Theory of Indistinguishables. A Search for Explanatory Principles below the Level of Physics.* 1981. (Forthcoming.)
151. J. C. Pitt, *Pictures, Images, and Conceptual Change. An Analysis of Wilfrid Sellars' Philosophy of Science.* 1981. (Forthcoming.)
152. R. Hilpinen (ed.), *New Studies in Deontic Logic.* 1981. (Forthcoming.)
153. C. Dilworth, *Scientific Progress. A Study Concerning the Nature of the Relation Between Successive Scientific Theories.* 1981.